Peter –
"There are more things in Heaven and Earth, than are dreamt of in your philosophy"
(v. Shakespeare)

Sean O'Donnell
Jany 20, 2008

THE PARANORMAL EXPLAINED

Intuitions and Time in Experience

SEAN O'DONNELL

First published in 2007

Copyright © 2007 by Sean O'Donnell

<antimemory.com>

PUBLISHED BY LULU

All rights reserved. No part of this publication may be reproduced, stored in a retrieval system, or transmitted in any form, or by any means, mechanical, electronic, electrostatic, recording, magnetic tape, photocopying or otherwise, without prior permission in writing from the publisher.

*"We must inevitably be living
with some scientific beliefs
that seem reasonable but are untrue,
simply because they have not yet been challenged..."*

Richard Milton - *Forbidden Science* - 1994

CONTENTS

INTRODUCTION		i
ONE – INTUITIONS ANALYSED		1
01	Coincidence Scrutinised	3
02	Intuitions Analysed	11
03	A Purely Temporal Phenomenon	19
04	Anti-Memory Describes	23
TWO – THE GREAT MISTAKES		35
05	Clairvoyance – the Muddled Inference	37
06	Telepathy – a Myth from Telegraphy	45
07	ESP – the Barren Oxymoron	55
08	Synchronicity – a Needless Mysticism	67
09	Remote-viewing – Clairvoyance Renamed	73
10	Dunne's Experiment with Time	85
11	Conclusions Confirmed	95
THREE – LEARNING TO PRE-CALL		107
12	How to 'Guess' Cards - Correctly!	109
13	3-Digit Pre-call	125

14	Roulette as a Game of Skill	133
15	Why it's Harder to Win!	143
16	Electrons Are Predictable!	153

FOUR – ANTI-MEMORY: WHY IT CAN BE   163

17	Does Time Really "Pass"?	165
18	Or Are We Just "Passing Through"?	171
19	Testing for Relativity	185
20	A New Kind of Consciousness	199
21	Questions Arising	211

APPENDIX – A   221

APPENDIX - B   225

NOTES   229

INDEX   235

INTRODUCTION

I've written this book to clear up those many confusions, for too long associated with intuitions of all sorts. Until now these confusions have caused intuitions to be regarded as paranormal, or beyond the capacity of normal science to explain.

Historically also intuitions have spawned many strange and irrational interpretations – implicit in terms like 'clairvoyance', 'telepathy', 'remote-viewing', 'synchronicity', 'ESP'. All of these presuppose mysterious mental powers supposed somehow to transgress space or distance, thus contradicting what the rest of science well knows.

By contrast in this book I will attain a very different conclusion – that intuitions are anomalies of just time alone. As such too they can readily be demystified, to the point where they don't conflict with conventional science at all.

Instead, as I will further show, they can contribute to normal science very usefully, and so extend it into regions previously unknown.

Paranormal events may of course embrace other possible manifestations, like psycho-kinesis (mind over matter), poltergeists, ghosts. By and large however intuitions comprise well over 95% of paranormal experience, and most of the formal research literature. So it's with them only that I'll be exclusively concerned.

To some extent also this work is a popularised re-write of my earlier book *Future-Memory and Time* (ISBN – 09528409 0 1), self-published by me ten years ago. While that book received moderate approval, and little significant criticism, it was generally felt to be somewhat too specialised for the average reader. My present offering is therefore an attempt to rectify this fault.

A great many people have contributed to the making of this book in various ways, and it would be now near impossible to name them all. Still my first thanks are due to solicitor Liam Gallagher of Galway, Ireland - for taking this topic seriously. Likewise to his brother Sean Gallagher, who must be the best-read physicist west of the Irish Sea!

I'm further indebted to Peter O'Toole from Moycullen - without whom this work could hardly have attained its present form.

My thanks are also due to Carol Craddock of New York, Bernadette Tierney and Colette Connolly of Galway – all of them helpful with various suggestions and ideas. I'm especially grateful to Brid Fagan from Barna and Michael Maycock from Spiddal, whose input was essential to the final manuscript.

More formally I'm indebted to Dr. Seamus O Grady and Prof. Markus Worner, at the National University of Ireland, Galway. They approved my annual course on "The Nature of Time", conducted throughout the pre-Millenium decade.

And finally my thanks to the ever courteous staff of The British Library at Euston Road, London. This is an institution unrivalled, for researching old facts to be recycled into new knowledge, as I am proposing here.

To *see* new truth, to *relate* it to the current state of knowledge, to *express* it appropriately - these are the three main tasks confronting any would-be advocate of new ideas. Of which in my case the last has proved hardest of all. By far, so far…

GALWAY
IRELAND
Oct 2, 2006

PART ONE

INTUITIONS ANALYSED

"The really valuable thing is intuition. The intuitive mind is a sacred gift, and the rational mind a faithful servant.

"We have created a society that honours the servant, and has forgotten the gift"

Albert Einstein (1879-1955)

1

COINCIDENCE SCRUTINISED

"Philosophy is often much embarrassed when she encounters certain facts which she dare not doubt, yet will not believe for fear of ridicule"
Immanuel Kant (1724-1804)

THE TELEPHONE RANG with a shrill insistence, its harsh tones disturbing the silence of my study room. They ripped through my half-formed thoughts on thermodynamics, which I'd been rather vainly trying to organise into a lecture for my class next day.

"That'll be Peter!" I said to myself with total certainty and yet irrationality.

My thought seemed certain because I "just knew" it would be Peter. This came to my mind without the slightest hesitancy or doubt.

Still it was also irrational because my errant friend had long been absent in London far away. He was not in the habit of phoning very much. Nor indeed at any particular time of day. Neither did I know of the slightest reason why he should ring me now.

And yet Peter it was at the other end of the line 500 miles away, his message merely that we could expect him back next week.

"It's another of those damn coincidences!" I said to myself in slightly exasperated mood………

I'd always been interested in the vagaries of coincidences from an early age. To me they seemed almost *paranormal* in their abundance: there were simply too many of them around for *normal* expectation to accept.

Too often therefore I would suddenly think of people just before unexpectedly meeting up with them. Other times I would encounter newspaper items on some exotic topic I had just been considering. Or I might hear old tunes play over the radio just after they'd come back to mind. Sometimes, as with Peter above, I would "just know" who was calling, before lifting the phone.

On several occasions during my undergraduate years I had even felt so curious that I had started to analyse these possibly paranormal incidents. Such immature efforts however seldom lasted long. For one thing those coincidences I encountered at that stage seemed to come and go like the seasons. Some weeks there might be many of them to observe.

Then there might be many months with none at all. However all that I'd really done so far had been to describe their essence critically:

> A coincidence happens when two similar but separate events, occur with apparent relation *around the same time.*

For example if some tree were to fall without obvious cause in your garden, and another in some distant friend's house next morning, undoubtedly that could be called a coincidence. Though hardly so if there was a 10-year interval between these two events!

But now in any case, after Peter's unexpected phone-call, I was much better equipped for this challenge of coincidence than before. With a good Ph.D. from the University of Edinburgh, I'd just completed a year in minerals prospecting as a geo-chemical analyst.

In practice this meant analysing rare metal traces found in ordinary clay. Its theory involved compiling significant patterns from weak anomalies (metallic over-abundance) as they emerged. Anomalies are violations of expectation, or departures from the normal rule, which demand explanation whenever they are found.

And now it occurred to me that my geo-chemical procedures could be transferred to researching coincidences. For the latter too were essentially weak anomalies, or perhaps paranormal deviations from normal expectation, which seem to emerge in common experience.

And finally I was living in Galway, Ireland. This was a small university city where people still had time to stop and stare and wonder, a way of life now largely vanished in the more frenetic 'Celtic Tiger' modern age. In fact life, as conducted in Galway back in the 1960s, was all very much a relaxed, slow and serene affair.

So that if coincidences were real phenomena, and not just chance artefacts of statistics, they seemed almost bound to flourish - *and further be noticed* - there if anywhere!

And so I decided to scrutinise the mystery of coincidence once again, but this time far more competently than before. My decision was based on little more than sheer curiosity. It was just the creative urge to tackle, a challenge of a totally unprecedented kind....

A *thorough* survey

Much later I would discover that remarkably few surveys of coincidence in any form had ever been carried out before. In fact there have probably been less than a dozen such ever conducted anywhere. And in any case few or none of these other efforts ever really resembled the venture on which I had now decided to engage. [1,2,3]

For most of those other coincidence surveys were really collected at a distance, or far from those who had reported them. Mostly they summarised incidents reported by strangers, events that had happened some time previously, and usually far away.

In contrast to this rather telescopic view of things, my new survey would be a much more microscopic one. For I was much closer in both space and time, to whatever new example might occur. Further I was determined that my new search for the meaning of coincidence should be as *thorough* as possible....

I started with some 16 friends whom I would quiz periodically about any coincidences they might have noticed recently. Some were postgraduate students, others were mostly professional people aged up to 40 years. Among them were Michael the artist and Morgan the hotelier, Nelly the nurse and George the doctor, Mary the beautician and Sean the barber, Eamonn the accountant and Frank the professor.

Some of these names I've altered slightly, but they will still be readily recognisable to those most concerned.

My search for coincidence was also never a proper census, nor even an amateur Gallup poll. I felt more like a botanist trying to pluck rare paranormal specimens, from the weeds of everyday normality. So that about once a fortnight or so – whenever I met them really – I would quiz these people about whether they had anything relevant to relate.

If so, I would quickly jot down any new tale of coincidence in whatever form, assuming these reports would be mostly reliable overall. Nor could I have any sensible idea of the bottom or cut-off point, below which there might be practical insignificance. So that I was always careful to record every last possible incident where coincidence just might, maybe, possibly, have been at work.

If therefore Jack merely thought of Jill, and then looked round to see her approach, that seemed to me worthy of note. Or if Mary reported she had been thinking of some melody that then came quickly over the radio, in went her account too. I was always determined to be totally thorough in collection here.

I also quantified each individual anecdote in categories familiar from geochemical analysis. So a seeming weak sequence – like Peter's telephone call above - might well be classified by me as 0.2 or just "interesting?" By this I meant that if there were 10 such incidents, ordinary sensible people might judge something anomalous behind at least some unspecified 2 of them.

More startling stories would then be graded as 0.4 ("relevant?"), 0.6 ("significant?"), or 0.8 ("highly significant?"). It all depended on just how startling, unusual or uncommon each individual coincidence might seem to me to be.

I had purposely made these gradings very low or unassuming when I began. Now I believe they were too conservative, much too conservative by far.

Incidents analysed

Most of my early records were indeed of a rather lowly kind. Typical was the following little 0.4 or "relevant?" anecdote:

> Thinking about the vast amounts of Guinness then being drunk around Galway, a bright idea suddenly occurred to Sean the Barber about noon.
> Perhaps the beer had some still unknown drug or ingredient in it, one which made people unusually partial to the magic brew?
> Within five minutes a friend cycled up to him, to propose the very same bright but bizarre idea!

As usual also, I analysed this little incident in traditional paranormal terms. That sheer *chance* might have been operating here was undeniable. On the other hand it did seem like a fair case for *telepathy* – paranormal communication between two distant minds. Conversely it could never have involved *clairvoyance* – paranormal knowledge of distant objects or places - because there was nothing physical to see.

Possibly too it might even have involved *precognition* – paranormal knowledge of some future happening. Though that seemed unlikely for such a trivial event.

Extra-Sensory Perception (ESP) seemed also a tempting possibility at first.[4] On mature reflection however I was forced to discard this notion as just a confusing oxymoron. For since perception means 'knowing through the senses', how could it ever be extra-sensory?

A somewhat firmer incident, which I assigned to the 0.6 or "significant?" category, seemed that reported by Maura, an arts graduate with a longstanding interest in coincidence:

> Maura was busy cleaning the bedrooms while her two-year-old toddler played happily downstairs.
> All of a sudden she felt an unprecedented urge to rush down and check on him. Only to find that her baby had somehow escaped from his playpen, and happily made his way outdoors.
> She caught up with him just as he was staring over the edge of the fast flowing stream, about 50 yards from her front door.

But Maura was not in the habit of rushing downstairs so precipitately. Sheer *chance* seemed therefore quite unlikely in this case. *Telepathy* seemed far more probable, though *clairvoyance* seemed also a possibility. And so too did *premonition,* traditionally supposed to come into operation whenever danger looms....

There were however other incidents where *telepathy* could be ruled out with some confidence. One such rare tale came from Mary the beautician, a sequence so bizarre and unusual that I immediately assigned it to the "highly significant?" or 0.8 category:

> Mary had been sent out by her mother, to do the weekly shopping with a £10 note. But when she got to the grocery shop, she realised that she'd somehow lost it on the way.
>
> Retracing her steps and finding nothing, Mary started to walk again. She dawdled along by a roundabout route, fearing the serious row to come.
>
> And there in the dust, by the old iron gate alongside the County Buildings, she came on another £10 note just like the one she'd lost earlier!
>
> Very sensibly she then proceeded to do the shopping as intended, not telling her mother about either incident!

This episode was also one of those few occasions which I investigated further, going up to examine that old iron gate at the County Buildings where the providential find was made. I found a small pile of rubbish blown by the wind into the corner alongside - cigarette packets, chip papers, yellowing fragments of old newspaper and so on. Likely then the money had been lost by someone with too much drink on board, attempting to stumble through this narrow gateway the night before!

Telepathy could hardly have been operating in this instance: nobody would knowingly see a 10-pound note lying in some rubbish and then just walk on thinking about the same. It might of course have been *precognition* – subconscious awareness by Mary that she would find just what she so badly needed if she walked this circuitous way.

But on balance *clairvoyance* – paranormal knowledge of a distant scene – then seemed by far more probable...

There were however other incidents where *clairvoyance* had very definitely to be ruled out. One such was a paranormal tale from Michael the intern, concerning events several years before. Still it seemed so peculiar that I had no hesitation in assigning it to the "highly significant?" or 0.8 category:

> Michael had dreamt that King Kong was chasing him through the Barna Woods. The beast was roaring mightily, throwing trees and caravans about him as he loped along. (The famous film had just been replaying locally.)
>
> About 10 days later however Hurricane Debbie unexpectedly crossed the Atlantic, wreaking large havoc along the Galway coast. And when Michael ventured out to experience Nature at its most furious, a strong sense of déjà vu began to worry him.
>
> For the constant roaring noise – roots cracking, trees falling, slates slipping – suggested he'd lived through all this previously in his dream!
>
> But it was only when he came to the Caravan Park in Salthill, that he found these suspicions were confirmed. For there were the mobile homes and caravans all splintered into matchwood, their remains thrown up by the storm against the Golf Course wall! Which finally convinced him that he'd somehow experienced this hurricane 10 days before!

Telepathy or *clairvoyance* could obviously be ruled out here. *Déjà vu* – that strong sense of having lived through some sequence previously – seemed the most ideal candidate. But then again it could have been *precognition* or even *premonition?* – forewarning of dire events to come?…

There are two kinds of coincidence

I ended my survey after 57 weeks with a total of 264 anecdotes, not counting my own purely personal experiences. Of this total I had assigned a mere 31 to the "significant?" (0.6), or else "highly significant?" (0.8) categories. On average, that worked out around one strong anecdote, from each respondent every six months or so.

But by then it had long been clear that there were really two very different types of coincidence involved. The first – and comparatively scarce – type of coincidence I termed *objective,* since it didn't involve any obvious anomaly of mind.

In fact there were only 7 such *objective* incidents reported to me – a mere 3% of my records. These stood apart from all the others because they had no clear anomalous input from mind. They might have been due to chance concurrence, though sometimes that seemed improbable. Alternately they might have been indicating some higher order of mystery:

> While driving out to Merlin Park on Monday, it suddenly struck Colman the broker that there was something familiar about the car in front.
>
> Then he realised that its number-plate had the same 3 digits (7 6 3), and in the same order, as his own. (Most Irish cars had 3-digit number-plates at that time.) Though of course the alphabetic letters didn't correspond.
>
> Driving back later a similar incident occurred. Once again the car in front had the number 7 6 3. Though this time it was a different vehicle!

Still people with cars will average perhaps 1,000 trips per year. And they will probably see at least 20 different cars in front of them every 3 miles or so. Also there were 12 regular drivers among the 16 people in my survey. So it was easy to calculate that one might expect such a report, about *two* coincident number-plates, from somebody in the group every few years. Assuming of course that all our drivers were as number-conscious as Colman.

But in any case I dismissed such incidents of *objective* coincidence from further consideration forthwith. There might indeed have been some kind of higher mystery behind them. If so however, it all seemed beyond my current competence.

The other 257 anecdotes of coincidence that I had collected (97% of the total), were all in a class very different to my 7 *objective* cases like that above. Instead they involved an alternative kind of coincidence that I termed *subjective*: these always involved some anomaly of Mind.

In this far more frequent type of coincidence, people would form a thought about something unusual – possibly through *telepathy* or whatever – only to encounter something very similar at some later point. Essentially also all such thoughts could be termed *intuitions* – examples of knowledge somehow attained without the conscious use of reasoning.

Most common 'coincidences' therefore seem to involve intuition in some form. Intuition I defined as the general faculty behind everyday intuitions, the plural denoting events where this faculty appeared to be involved.[5]

257 intuitions recounted by 16 people over 57 weeks also meant an average of 1 potential intuition per person about every month or so. However this average was never at all distributed evenly. For example I observed a striking increase in anecdotes as the weeks went on. So that there were just 89 cases in the first six months of my survey - and then 168 in the second half-year.

Here the first possibility was obviously that people were only fooling themselves (and me!), seeing ever more insubstantial shadows in their everyday experience. But later it seemed more reasonable to conclude that my people were really growing more intuitive - or perhaps just more observant - as the days went by.

For one thing more ordinary experience will support this second possibility. For example if you develop a new interest in flowers, or clouds, or pop tunes or whatever, then you will be likely to notice ever more frequent examples in your daily routines. Most of these you would have seen - but never really observed - previously. Or in other words you would have paid little conscious attention to them all.

Neither were all my respondents equally fruitful in their reports. For example 3 people never reported even one single intuition or coincidence between them, over the entire yearlong exercise. Here it seemed simplest to suspect that these people were not very observant people in this respect. Or perhaps not really much interested in coincidence at all.

Of the others, just 4 people reported 124 incidents between them, or nearly half of the total logged. Most prominent among these was Mary the beautician (she of the ten-pound note anecdote above), who had a reputation for intuition among her peers. It was Mary too who made one of the most perceptive remarks of anyone, an observation from which much progress would result eventually:

> "There are days when I notice lots of them (intuitions) – and then weeks on end when I get none at all."

And so there arose my concept of the *psi-state* – a rare, evanescent and intuitive state of mind. Psi is a noun introduced by Cambridge psychologist Robert Thouless. It denotes those mental processes apparently associated with 'psychic' events.[6]

That "words are the carriers of thought" is a maxim well known to linguists and philosophers.[7] This is a very important matter in science, and one to which I will return again. Meantime this particular neologism I'd just run together, would eventually prove a very fruitful one.

Soon afterwards therefore I started on my Second Coincidence Survey. But this one would be concerned with my own purely personal experience only, and it would focus on just intuitions alone….

A brief summary

From my first Coincidence Survey I concluded that there are two very different kinds of coincidence in everyday affairs. By far the more common type – 95% of the total – could be described as subjective intuitions, and they were not evenly distributed among the populace.

Intuitions seemed to flourish as one grew more interested, and their essence was always some strange thought reflected in later experience. This linkage might then be interpreted by various common paranormal terms like *déjà vu, premonition, precognition, clairvoyance, telepathy*. Though whether all these terms were really valid was still quite unclear.

2

INTUITIONS ANALYSED

> *"The first principle is that you must not fool yourself, and you're the easiest person to fool"*
> Richard Feynman in "Genius" by James Gleick (1992)

IN DECIDING HENCEFORTH to analyse only my own personal intuitions, I had cut out all further consideration of *objective* coincidences, which after all just happen occasionally. And with this focus, on purely personal experience which looked potentially paranormal, I was again striving to get as close in to the facts as possible.

Instead of mere *second-hand* reports furnished by others, surely primary or *first-hand* experience should yield even more insights than before?

By now I had glimpsed that intuitions could perhaps maximise, while people are in a rather special state of mind. This special psi-state I might first learn to recognise, then strive to duplicate, and finally reproduce deliberately by controlled exercise of will.

A more observant state of mind

I began by collecting together those various cases of personal experience I had recorded up to then. There were 47 such incidents logged in my diaries over the past 5 years. Of these over half (27) had been logged during the past year, when I was surveying others about their anecdotes of coincidence. I was then apparently growing more intuitive, or at any rate more observant of intuitions, as my interest in them grew.

Sceptics or strict behaviourists might of course decry a certain loss of objectivity, in this new procedure of mine. Indeed a total cynic might even cite *identifying paramnesia* – a form of confusion about sequences in time.[1]

Still soon I discovered that dangers like these were all more apparent than real. For now that I enjoyed direct and immediate access to such facts as were involved, my approach was even closer or more microscopic than before. Likewise I could have every confidence that I did indeed *know* what I was talking about. And *know* in the most literal sense too!

In addition I was determined that my Second Survey of Coincidence – which had now evolved into a deliberate analysis of intuitions - should be even more thorough than the previous one. So for each separate incident I recorded some 13 possible background variables initially, factors which just might conceivably be relevant.

Co-factors like the weather, moon phase, compass direction, distance and time lag (between thought and outcome) – all these and others I therefore started recording for each separate incident. Gradually however I was able to discard most of these early variables as irrelevant. So that from those 13 co-factors I used to record initially, I could find only 3 of much significance in the end.

These 3 co-factors were time-lag, degree of interest, and my background mind-state at the time.

Concerning this last I noticed early on that my observed intuitions seemed to recur in clusters, just as observant Mary had remarked originally (Ch.1). So that sometimes I might notice a half-dozen incidents within a few days. And then find nothing at all to record for several weeks afterwards.

Clearly therefore my own intuitions didn't recur on anything like an average or regular basis as the year went by. They were not at all evenly distributed across my weeks and days!

But once this uneven distribution of personal intuitions had been clarified to my satisfaction, a second factor gradually grew more obvious. Now I could see that their varying frequency was broadly correlated, with my general or background state of mind.

If therefore I felt unusually harassed, rushed, harried or busy, I had very few cases of likely intuitions to record. These are all states of experience we might now describe as having 'high mind-noise'. So quite possibly I was just too preoccupied by other matters to notice intuitions as they occurred!

Conversely there were other times when one felt more relaxed, or more at ease with the world at large. During these states of 'low mind-noise', I would usually observe far more and stronger cases of intuition to occur. This was all very much as observant Mary had stated previously. And indeed as numerous gurus and mystics have told us down the centuries.

Very gradually then I began to recognize a certain feeling of *presentiment* before my more outstanding intuitions came to be. I could know more readily which of my thoughts were likely to be complemented, as intuition at some point afterwards.

Clear observation of this factor then led to a further procedural modification, one that began to pay off almost immediately. Once any apparent intuition had been confirmed by some coincidence in the outer world, I tried to think back to that mind-state wherein the original thought materialised.

For example was my mind a blank or otherwise? What other mental circumstances were apparent when it formed?

Much later I found out that such procedures come under the general heading of *phenomenology.* This is a branch of psychology- cum-philosophy pioneered by Edmund Husserl (1859-1938) a century ago. Among other things it involves examination and description of internal mind states.[2]

But for me at that time in any case, the important thing was that my own unwitting brand of phenomenology appeared to work. So that, given a favourable background, I could "steady" myself into a general state of passivity, concentrate gently on holding it - and then latch on to whatever thought came first to mind.

Increasingly such thoughts proved to be recognisable as significant through presentiment. And afterwards complemented by coincidence in the outside world.

For the benefit of those readers who may wish to emulate this feat on their own, I've included instructions on "How to Grow More Intuitive" in Appendix A/. Though really these instructions differ little from what mediums and mystics have been saying for centuries. They're also a version of what you can now find readily in any good self-help intuition book.

But in any case for me back then, it was clear that my previously ignored intuition faculty, was indeed growing stronger by the day...

Previous pioneers

At the time, and because I had only the faintest wisps of hints for guidance, this all seemed to me a rather impressive learning feat. Only later would I come to realise that some others, together with numerous mediums and mystics, had long been saying and doing much the same.

For example, as we'll see in Chapter 10, J.W. Dunne in the 1920s had advocated a similar approach. There was also Gilbert Murray (Ch. 6), a British classicist who had developed his own learned intuition skills. So too had those several dozen participants in the CIA's remote-viewing program (Ch.9), which more or less proved that anyone can do the same.

But one of the best ever descriptions, of learned intuition, is given by American author Upton Sinclair in 1934. He wrote a book wherein his wife Mary Craig describes her own 'telepathic' feats.[3] She was a self-taught intuitor, all very much as I was now trying to be. So later (Ch. 6) we can consider her recommendations in more detail.

People like these all agree that to attain the intuitive mind-state in such ways is "no easy matter" at first. One part of the mind seems to rebel against deep self-liberation of this sort. A similar and unceasing internal struggle with the interfering Ego, is the main point emphasised in Timothy Gallwey's various *Inner Game* books.[4]

But in any case I soon got accustomed to this initially strange exercise. I could then settle down to attain intuitions more or less at will. I was travelling a road which others had often trod before. I was perhaps unique only in that I combined such direct experience with 'hard' science skills.

I might therefore perhaps progress a little more readily, along that straight path of due logic only, which clear thought requires.

More quantity - and quality

From 349 personal intuitions eventually observed in this way, I judged the majority as of decidedly low quality indeed. Though usually that little bit bizarre in their content, they still seemed notably trivial, banal, mundane. So I consigned the majority of them - 181 cases or 53% of the total - to the lowest "interesting?" or 0.2 category. And yet by the end I'd learned more from these, than from those other more spectacular ones:

> Being in an especially good mood on this sunny Monday morning, I started to hum the old half-forgotten English tune "Greensleeves", which seemed to me to reflect the spirit of the day.
>
> Within a few minutes, the same tune started over the radio! But perhaps the disc jockey was just animated by the same sunny morning as myself?

Clairvoyance? Telepathy? Precognition? Chance? Causality? All options, both normal and paranormal, seemed equally feasible here.

But in any case soon afterwards I deliberately set out to organise my life-style as already described. I would minimise all those common daily irritations to attain lowered levels of 'mind noise'. All bills were paid on the day I got them, all letters answered likewise, all irksome demands on my attention suitably minimised!

The result was a large and gratifying increase of personal potential intuitions, soon rising markedly above their previous frequency. So that I recorded 47 potential intuitions over 5 years before starting, 118 during the first half of this investigation, and 184 in the second 30 weeks. Not only that, but the later intuitions were of rising quality. Further I could better detect that slight hint of presentiment, when some new incident was due.

The following "relevant?" or 0.4 incident was typical:

> Sunday 11.15 am: In an idle moment I decided to leaf through my phone diary, deleting those numbers no longer relevant. I came to one Taffy Howe, from Glasgow about 250 miles away.
> Since I hadn't heard from him in many months, I was deliberating briefly whether to cancel him out or no.
> Whereupon, almost immediately, Taffy rang up quite unexpectedly!

Now what strange influence could have impelled Taffy to pick up the phone just when I was about to write him off, and after a year of no contact? Or was I reading this case correctly at all?

As always when each singular case like this was considered in isolation, it could still be easily explained away as chance coincidence. But considered *en masse* or collectively, a very different conclusion was becoming unavoidable. I felt I was closing in on – or more likely opening up - some strange, mysterious and novel capability of mind.

Later I found that Freud had referred to a feeling of being lost or powerless, when one is repeatedly confronted with unlikely events like these. Such happenings we often term –

> "uncanny ...... a class of terrifying which leads back to something long known to us, once very familiar". [5]

All of which seems to me to make eminent sense, where strong intuitions are concerned. Was I therefore discovering some wholly new and vaguely threatening process here? No! It felt far more like uncovering something always present but previously ignored:

> Today I somehow fell to thinking about an intelligence test presented to children in America: how to retrieve a coin which has fallen down through a pavement grille. The supposed solution is to lower down chewing-gum on a string, so that it can cling to and retrieve the coin.

> But now I suspected this wouldn't work in practice – through lack of contact pressure between the two solids involved. A better method might be to jam some gum on the end of a stout stick, then pressing it down through the grille so that the coin below it could adhere.
>
> Later that evening I was passing the Eyre Square bus stop beside the old Hibernian Bar. There I chanced to glance down through the cellar access grille alongside – only to spot a half-crown lying on the floor!
>
> The point is that half-crowns were significant money at that time. So that this valuable coin can hardly have been lying there too long.

Now I was no more in the habit of thinking about coins lost under a grille than the next person. Neither could I recall ever having observed any such, at least within the past several years. So I had no hesitation in assigning this potential intuition to the 0.6 or "significant?" category.

Clairvoyance seemed the most likely paranormal possibility, since it involved knowledge of a distant place? *Telepathy* seemed also possible, perhaps from some stranger who'd seen the coin like I did before passing by? But then too it could have been *Precognition* on my own part – an earlier inkling of a later scene I would soon meet?

Now too I began to sense that my long sought common pattern for personal coincidence – which I'd been seeking so relentlessly over the past few years – was almost within reach. But still it lay tantalisingly just outside the grasp of intellect. I felt as if it were hidden behind some flimsy veil of conscious inhibition – one I could almost, but not quite, see through.

The elusive butterfly

Eventually all this process of puzzlement, culminated in one weird but "highly significant?" (0.8) incident. Through it all my many confusions suddenly clarified:

> Saturday- 4.30 pm: Today around 2 o'clock it suddenly occurred to me, that my pursuit of intuition was all very much as described in 'The Elusive Butterfly'. This was the name of a highly metaphysical pop song one that had been near top of the charts for several weeks:
>
>> "It's only me pursuing something I'm not sure of,
>> Across my dreams, with nets of wonder,
>> I chase the bright elusive butterfly of love…"
>
> Now suddenly the similarity – between these words and my own pursuit of intuitions - seemed nearly total, perfect and complete. And yet this similarity had never occurred to me previously, despite my having heard this song so often over the past few weeks.....

Just over an hour later I went on a casual stroll up the busy thoroughfare of Prospect Hill, a most unusual route for me.

And there I observed an excited man in gum boots, weaving down on the road through the passing traffic, as he tried to catch a lone white butterfly that kept fluttering just above his reach!

I stared in amazement at this scene for about 20 seconds, which seemed more like an hour to me. Until finally the butterfly soared up out of reach into a clear blue sky, on a thermal from the iron roof of the old Leather Shop nearby...

Later I learned that this eager lepidopterist was a well-known local 'character', whom we would now call autistic, much given to clasping his hands and leaping about excitably. But nobody could ever recall him chasing butterflies before! [6]

In any case I had no hesitation in assigning this sequence to the "highly significant?" or 0.8 category of intuitions. Of these I'd only ever observed some dozen to that point.

Still in addition there seemed to me something almost supernatural about this strange episode. It might almost have seemed to suggest some hidden link with Nature, as with Jung's somewhat similar Tale of the Golden Scarab (Ch.8). If so, it could be indicative of some higher order of mystery quite beyond my present competence.

But in any case I'd never before seen any adult chasing butterflies anywhere - much less down a busy main street. Though as my ever critical friend Kerrigan now pointed out, perhaps its pursuer had been inspired, by that same song as me!

In the end I decided to file this entire incident apart, as perhaps indicating some higher order of mystery. Along with some half-dozen other incidents in total, it seemed to raise questions too complex for my current capabilities.

But meantime I could only continue with my more mundane analysis as before. Mere *chance* concurrence – between my earlier thought and later encounter – then seemed to me most unlikely here. In addition both *telepathy* and *clairvoyance* were obviously ruled out.

Which also confronted me with just one possibility - stark, unavoidable and clear. *Precognition* was the only possible paranormal inference here!

And with this fact definitely established in this instance, a huge clarifying simplification suddenly came of a rush to me. It now seemed so obvious that I could only wonder – and later investigate - why it had eluded me for so long. Then I turned back to my records to check them all anew…

A brief summary

First-hand experience showed me that intuitions do indeed grow more frequent as one becomes more interested. Further they can be made to flourish by deliberate emphasis on 'low-noise' mind states. And finally *precognition* seemed to be the only paranormal inference required.

3

A PURELY TEMPORAL PHENOMENON!

"Mysteries must not be multiplied beyond necessity!"
William of Ockham, d. 1349

I NOW SET TO WORK on my records with a will. No miner nearing a gold deposit could ever have dug through dross more feverishly. No Fraud Squad detective in pursuit of some business suspect, could ever have leafed through his records more assiduously.

I wanted to see if *precognition* was the one single process required in all my observations so far!

Of those 349 purely personal records I had accumulated, there were 12 that simply made no sense at all. Neither *telepathy,* nor *clairvoyance,* nor *precognition* – nor any possible combination between them – could provide any viable interpretation for these. While sheer *chance* still seemed unlikely at most times.

So I set these few cases aside as a higher order of mystery, secure in my understanding that pioneering research can seldom deal with everything in any new field initially.

Telepathy	*Clairvoyance*	*Precognition*	*Total*
+	(+)	+	253
–	+	+	046
–	–	+	<u>038</u>
			337

TABLE 3.1: *Precognition* **was the only factor required to interpret all intuitions personally observed.**

For those remaining 337 cases I had on record, 253 might have been ascribed equally to either *telepathy* or *precognition* - and occasionally to *clairvoyance* if one were to stretch things. 46 could only be ascribed to either *clairvoyance* or *precognition* on their own. This still left 38 incidents that could only have been ascribed to pure *precognition* alone.

For clarity I then arranged these results in a little table as above. This showed very clearly that *precognition* was the only required paranormal factor, one needed to invoke throughout.

My next step was entirely obvious. Would similar results obtain with that First Survey of Coincidence as described in Ch. 1? I hurried back to their records with still more eagerness....

As I've already mentioned about this First Survey, there were 7 of those 264 records that had to be excluded as very different from the rest. A few of these seemed best interpreted as a higher order of mystery. The rest could reasonably be ascribed to *objective* chance coincidence. Which then left me with 257 cases, for tabulation as before:

Telepathy	*Clairvoyance*	*Precognition*	*Total*
+	(+)	+	195
−	+	+	036
	−	+	026
			257

Table 3.2: In my earlier Survey of Coincidence also, *precognition* was the sole interpretation required.

Now, as any statistician may notice, agreement between the percentages, in both tables and for all 3 classes, is unusually close here. This might have been due to unconscious bias of interpretation on my part. For example could *telepathy* really be favoured over *clairvoyance* in some particular incident?

Total observations recorded =	349	+ 264	=	613
Not analysed =	12	+ 7	=	19
Total accounted for by *precognition*			=	594
Percent accounted for by *precognition*			=	97%

TABLE 3.3: 97% of everyday coincidences observed could be interpreted, in terms of *precognition* alone.

But in any case I decided that this was again a secondary effect or maybe a higher order of mystery. And so it seemed best set aside as before. Perhaps it might be explored by others in more depth some day.

Ockham rules!

Meantime my tabulations now clarified unavoidably, that common pattern I'd been seeking for so long. For almost all those 600 intuition anecdotes I'd analysed to date, *precognition* was the one, sole, and only paranormal process required, This was a clearly unavoidable fact henceforth. Presumably I'd been too timid or too stupid, perhaps too conventional - or maybe even subconsciously fearful? - to see it properly before.

More accurately also my new common pattern could be described, as some mysterious process of transcending *time alone*. This was that elusive butterfly of common pattern I'd long sought so vainly, but always too blind to see the obvious. Yet now that it appeared so very clear and inescapable, I couldn't but wonder, at just why I had missed it for so long!

Perhaps this was because it all seemed such a huge transgression, of so much that I'd previously taken for granted, about the role of time in everyday affairs?

But in any case what of those other alternative paranormal notions - like *clairvoyance* and *telepathy* – where *space* transgression had formerly seemed operational to me? Why not consider them also in those many cases where they'd once seemed so feasible?

To this the simple answer was there was no logical need to consider them anymore. For they'd been rendered redundant by an ancient law known as Ockham's Razor. It had its origins in the murky world of mediaeval theology. Yet amazingly it still survives as one of the most basic rules in all modern science methodology.

We owe this venerable and very useful rule to one William from Ockham - a small Surrey village southwest of London during mediaeval times. William was a Franciscan monk and logician, a man who won many intellectual battles in his day. Born around 1285 at Ockham, he probably died of the Black Death, in Munich in 1349.

In those days people were often named by where they came from. So when William formulated his powerful new rule of logic it became known as Ockham's Razor. This was because it could cut so easily through those many tangled theological disputes of that era.

Exactly what William said originally (in Latin of course!) now seems lost to posterity. But it was something like *"Mysteries must not be multiplied beyond necessity."* Or in other words you needn't consider several explanations where just one will do…

When science then got really going in the centuries after Ockham, his logic soon proved invaluable. This was because it stressed the great value of simplicity when confronting mystery. So that today it's accepted almost automatically, part of the unwritten methodology of all scientists everywhere.

At this point in my analysis of intuitions, I was therefore left with little scientific choice. There was simply no further need for me to consider all those other traditional notions - like *clairvoyance* or *telepathy* - any more. For that clear common factor of *precognition* had now rendered both of them redundant, it being the one single paranormal interpretation now required!

Or if indeed there were any of those other manifestations in reality, they should logically be left aside for the moment, as perhaps indicating some higher order of mystery!

But whence had all these other traditional paranormal notions come? And would my purely personal conclusions still hold in wider situations everywhere? Could *precognition,* or perhaps something similar, really be the only option ever required for intuitions of all sorts? And was it further really the best term, for all that I'd observed so far?

A brief summary

Precognition alone could indeed account for 97%, of those 600 potential intuitions I had analysed. No other paranormal option was then logically required. But would my pure personal conclusions still hold elsewhere? And whence had all those other paranormal notions originally come from?

4

ANTI-MEMORY DESCRIBES

"It's a poor sort of memory that only works backwards!"
Lewis Carroll – *Through The Looking Glass* –1871

BEFORE EXAMINING the utility of that term *'precognition'* however, we need to clarify what intuitions actually are. This requires a wholly new redefinition, one which can encompass their facts with more accuracy than anything in the dictionary.

> An intuition is some unusual thought or idea – which *must be confirmed* by observation at some later time.

Here the second half of this new definition is by far the most essential point. It's also one universally overlooked before.

Again too this point is best shown by a little illustrative anecdote. Consider therefore the following early incident from my own personal Intuitions Survey - one which I'd assigned to the lowest 0.2 or just "interesting?" class:

> Thursday, 11.36 a.m.: Molloy and I were strolling up past the new Post Office, when I suddenly remarked that we hadn't met up with Nancy recently. What made me remark this I just don't know. For she was merely a casual acquaintance on the social scene.
> About 1 minute and 100 yards later however we rounded the corner by the restaurant on our way up to the Square. Whereupon we literally bumped into the aforesaid lady coming down the street!
> Molloy dismissed this encounter with the single word "Coincidence!", and kept on boring me with his hackneyed Space Race ideas.....

Supposing however that I'd merely remarked on Nancy - but then never encountered her so quickly afterwards. In that case I would merely have uttered a casual remark of no great consequence, one likely to be forgotten almost immediately! There would have been nothing to hint that intuition might have been at work.

Importantly therefore paranormal intuition must always require confirmation through ordinary observation later on. Though their correspondence with such later confirmation is seldom totally correct in all detail. All of which now left me with some tentative conclusions about the mystery overall:-

1/ Intuitions are associated with a rather special state of mind.

2/ They're just strange thoughts which seem to involve foreknowledge of some future experience.

3/ Of necessity they must always be confirmed, by ordinaryor common observation later on. Though their correspondence with such later confirmation is seldom totally correct in full detail.

4/ Most 'trivial' intuitions reflect 'slight' events in the immediate future - a matter of minutes ahead. Whereas more striking intuitions tend to reflect more interesting experiences – which may be encountered only days or weeks afterwards.

5/ The longest such interval I ever observed concerned a strong intuition of a drowning - which I then personally encountered 3 months later on the Grattan Promenade, just 130 yards from my own front door!

de Pablos on dreams

This latter pattern, of exponential decrease over time for intuitions, is also one that has since been well confirmed. Of these later confirmations by researchers into the paranormal, the most recent is that by F. de Pablos who has analysed his own dream intuitions as complemented afterwards.[1]

Of 124 such intuitions, de Pablos finds 75 confirmed during the first 24 hours, 17 during the next day, and 19 over the following three weeks. There were also 6 intuitions apparently complemented after 1 month or more. All of which affords a clear pattern of exponential or logarithmic decrease with interval.

In any case my own conclusions up to this point now left me with the further rational requirement to explain! To *explain* is to "make more sense" of former mystery. This necessarily requires that you compare it, to something else which you already know.

Reverting then to my new understanding of intuitions as purely temporal phenomena, the essence of their mystery slowly clarified. To explain them one must seek out something similar, from other and more familiar mental activities or mentalities.

The analogue of memory

The obvious, indeed only, candidate here is memory. Through this familiar mentality we can somehow create some semblance of our past experience. Though how we can do this is very much a mystery. For the mechanisms of memory are still totally unknown.

Nobody therefore yet understands how you can now recall details from your distant childhood. Nor indeed even how you can now remember, what you had for dinner an hour ago. And even though memory does somehow inform us of such *past* experience, it seldom does so with full accuracy.

In any case intuitions appear to inform us occasionally, about *future* happenings likewise. To a first approximation anyway, they can then be described as memories that seem somehow inverted in time.

In addition we regard memory as normal because it's such a well-known and familiar activity. Conversely we regard intuitions as paranormal - beyond normal explanation - because they're much more infrequent or unfamiliar. Further they seem to conflict with our sense of time, and what we like to think that we know about the same.

Or in short we regard memory as a *normal* operation because we use it so frequently, and always with a definite orientation towards the past.

Conversely we regard intuitions as *paranormal* because we experience them so infrequently, and further because they seem so curiously oriented into future time.

All in all therefore intuitions now "start to make more sense" as a sort of opposite to memory. Their sole anomaly is just that they seem to be a time-inverted version, of the more usual process.

However this interpretation I've just developed is by no means a totally new idea of mine. Indeed it goes back at least to the redoubtable St. Augustine, who died as the Vandals were hammering at his city gates in 430 AD. A famed early Christian philosopher, he's now also recognised as a premier psychologist of time:[2]

> "Those things which we sense do not enter the memory themselves, but their images are there ready to present themselves to our thoughts when we recall them...
>
> "Whether some similar process enables the future to be seen, some process by which events not yet occurred become present to us by means of already existing images of them, I confess, my God, that I do not know..."

Later philosophers have also touched on the same theme, particularly over the past century. For example Bertrand Russell, the doyen of all British philosophers, thought it 'merely an accident' that memory works backward but not forward - so revealing your past but not future experience:[3]

> "So our relations between Past and Future would be symmetrical were it not for some fortuitous quirk of mind."

Another commentator who thought likewise was Alfred Jules Ayer (1910-89), premier British philosopher of the mid-20th century.[4]

> "It's assumed that any evidence which tends to show that precognition does occur can be rejected out of hand, or at least that it must somehow be explained away....
>
> "Yet there is no a priori reason why people should not succeed in making true statements about the future in the same spontaneous way as they succeed, by what is called the exercise of memory, in making true statements about the past...
>
> "The argument against precognition is, therefore, not logical but empirical; the evidence in favour of its occurrence is still very weak."

Ayer's contemporary C.D.Broad (1887-1971) also highlighted the analogy of memory:[5]

> "The fact is that people who have tried to theorise about non-inferential precognition have made two needless difficulties for themselves by making two mistakes.
>
> "In the first place they have tried to assimilate iot to sense-perception, when they ought to have assimilated it to memory."

The Prime Assumption

From the viewpoint of personal experience and deeper analysis however, such philosophical conclusions embody four large defects. For a start they all rest on a huge but untested assumption, one apparently never realised before. I'll term this *The Prime Assumption* because it's such a fundamentally important one:

> *The Prime Assumption* is merely that Mind or memory, must of necessity be limited, to direct contemplation of the past alone.

Strangely nobody ever seems to have articulated, must less investigated this matter, before I first expressed it in 1973.[5]

And in formulating it then for the very first time, I also left out the physical present for good reasons. It's already long past in reality by the time we become aware of it! For example if somebody happens to stamp on your big toe while you are gazing elsewhere, it will probably be at least 20 milliseconds later, before you realise the same![6]

In any case all philosophers of time (including those above) have always just simply *assumed,* that Mind or memory must inherently be limited, to your past experience alone. But without ever making the slightest effort to validate or check up on, whether this fundamental belief of theirs was really true.

And yet, as the common experience of intuitions suggests, in reality things may well be otherwise.

Meantime vast realms of thought, logic and philosophy, all rest on this wholly untested *Prime Assumption* or belief. So that if it were ever invalidated or falsified, much of their basic foundations would be just so much intellectual sand!

In science or logic also, unchecked assumptions of any sort are always necessarily suspect. For one thing too many of them have proved wrong before. Such false beliefs as the sun's daily traverse across the heavens, or the assumed immobility of continents, once seemed so obvious that few ever thought to check on them at all!

Often too this was because the consequences – like an Earth spinning at 1,000 miles per minute round the Solar System – seemed far too problematical to contemplate.

For the *Prime Assumption* also, something similar now holds. It has always served like a negative catalyst. It was a great blocking agent confusing all clear thought, retarding much further progress in the general field of subjective time…

Secondly in any case, none of these whom I've quoted above, ever apparently had much personal experience of intuitions from which to generalise. Which must have left them rather like "blind men probing rainbows" in this context.

Neither did any of them ever make much, indeed any, attempt to progress beyond the broad and somewhat vague idea of memory in general. They never considered precisely which memory sector – recognition? recall? retention? - intuition might resemble most.

In turn this led on to a final operational mistake. They accepted that word '*precognition*' without question – i.e. with no examination of how adequate or descriptive this term might be!

That "words are the carriers of thought" has however long been realised, as indeed I've mentioned earlier.[7] From which too there follows an important corollary, one for which science history can afford abundant examples: *Wrong words can inspire wrong thoughts and so further wrong deeds*. Or as George Orwell said: [8]

> "(Language) becomes … inaccurate because our thoughts are foolish, but the sloppiness of our language makes it easier for us to have foolish thoughts."

Anti-memory best *describes*

All in all therefore it seems wiser to replace that word 'precognition' by anti-memory, a term I first proposed in 1985.[9] In science the prefix 'anti-' is commonly employed, to denote properties opposite to those more usually observed.[10] For example physics has long known of anti-matter, an exotic material with electric charges all opposite to those in matter of the more familiar kind.

Anti-memory therefore treats intuition, as a paranormal memory of some incident *before* the actuality is observed, and not *afterwards* as in the normal case. Its great advantage is that it permits intuitions to be considered in a new, open-ended, and more productive way.

For example if intuitions are now considered in terms of anti-memory, they need no longer seem so totally out-of-this-world or mysterious, with absolutely nothing similar for comparison. Instead we've now got a 'handle' on those rare events which can make them more tangible. So that they begin to "make more sense" than before.

In my own case this also allowed better understanding, of almost all those 600-odd cases of intuition I'd analysed so far. Further it encouraged consideration, of those various similarities and dissimilarities, to the more common memory experience.

I therefore decided that I would describe all intuitions in terms of anti-memory henceforth….

Pre-call is more precise

This also meant that I would discard that old term *precognition*, a word introduced for some kinds of intuition by F.W. Myers around 1894. He was a co-founder of the Society for Psychical Research (SPR), in London a dozen years previously. [11]

Though Myers' term *precognition* had long been used in theology, he adopted it into the lexicon of the paranormal very casually indeed. He reported no apparent tests for its fitness, no deep consideration of how well it might "fit the facts" of intuition as readily observed.

Obviously therefore this term 'precognition' requires critical examination all round. Or is it at all a valid description, for intuitions as they actually do occur in reality?

Here firstly that common word 'memory' has long been realised as a broad or blanket term, one covering several more specific aspects of Mind. Of these only 'recognition' and 'recall' need concern us here.

The point is that recall requires just 2 essential elements or components - while recognition is more complex and needs 3. This all-important difference can also be shown by two brief anecdotes. Consider then first the following short sequence:

> "I met Mary yesterday. Now I can re-call (Lit. - summon back to mind again), some details of that green coat, which she was wearing then."

This story clarifies how there are just 2 essential elements to any re-call sequence. First comes an experience at some point in the past. Second there is a conscious effort that you make later, to summon something of that experience back to mind.

Re-cognition in contrast is a more complex phenomenon. It has 3 - instead of just 2 - essential elements. Again you can see this from another short sequence:

> "I met Mary yesterday. I see her walking towards me now again. I re-cognise her (Lit. - know her again) because my mind somehow connects these two events."

For re-cognition the 3 essential elements are then a past experience, a present experience, and a rather mysterious act of thought by which we bridge the two.

Reconsidering therefore intuitions in terms of 'anti-memory', we can usefully recount a very common type of anecdote:

> "Joan suddenly thought of long lost Jill for no good reason. Whereupon the latter rang up shortly afterwards!"

This clarifies how intuitions have just 2 essential elements - a prior thought and a later experience. So that re-call - but not re-cognition! - is that memory function, to which they best correspond.

Arising from which intuitions are most closely described in terms of *pre-call* – a neologism I first proposed in 1973. To *re-call* is to summon to mind again or afterwards in the normal way. To *pre-call* would be to summon to mind beforehand more paranormally. [12]

From which it must also follow that *'pre-cognition'* – a word long accepted by paranormal researchers but without even the slightest reality check - is certainly not an accurate descriptive term. It's confused and mistaken because it must inevitably suggest an unspoken similarity with *'re-cognition'*. Even though in fact there's no such similarity at all!

All of which therefore explains why I decided to banish *precognition* from my operational terms, my resultant thinking, and indeed this book you're now reading, from that point on.

Henceforth I would use either anti-memory or pre-call as required, the two terms being largely interchangeable.

This new decision of mine seemed yet another advance, in making the complex even simpler than before. For it reduced that huge and bewildering variety, of all those varied intuition cases I'd analysed, down to a less demanding state of more simplicity.

In fact it now only required that single letter 'p', prefixed before that common term *re-call,* to wrap up all these intuitions, within a radical new idea! This also seemed to me a good example of what science is really all about - the reduction of diversity in reality to identity in knowledge, through minimalism relentlessly pursued!

Description by diagram

In science further it often helps if you can produce a visible diagram of your idea. For the Ancient Greeks a diagram meant literally "drawing through" some topic, so giving a little picture meant to encapsulate the essence of the whole. This also can help your thinking greatly, since it brings an extra factor of visual intelligence into play.

For example most people would find geometry immensely more difficult without all those diagrams of triangles, circles and so on. So too would chemistry be without the familiar diagram of the Periodic Table. Or biochemistry with no spiral diagrams of DNA.

Finally the same clarification happened with Einstein's first Theory of Relativity, which was first expressed through algebra. It only gained wider attention after H. Minkowski turned the equations, into more visual space-time diagrams (See Ch. 18).

At a very early stage therefore I was therefore happy to discover that my new concept of anti-memory or pre-call could be expressed in visual form. This required just 4 simple steps or increments, to build up into a single explanatory diagram.

ANTI-MEMORY DESCRIBES

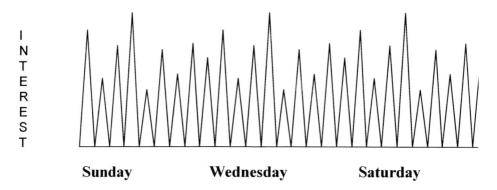

Sunday **Wednesday** **Saturday**

DIAG. 4-1: Jan had many experiences of varying interest throughout last week.

First one can diagram the fluctuating levels of interest or attention, of a young lady I'll call Jan, as she lived through last week. High peaks denote observations or events of outstanding interest to her, while low peaks denote less interesting ones..

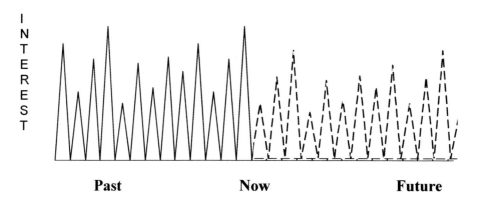

Past **Now** **Future**

DIAG. 4-2: At midweek, last Wednesday noon, Jan's past experiences were on a different conscious status to her future ones

Second we can consider Jan at midweek, which was noon Wednesday. There's now an apparent great difference between the two halves of her week. To the left are all those past encounters she has experienced already. To the right are all those other future happenings still not experienced and to come. This difference we can define by showing Jan's past in suitably solid or definite outlines, her future in more broken or less definite ones

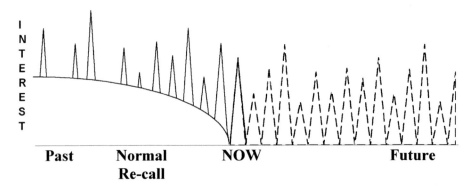

DIAG. 4-3: A conventional, or past-oriented, memory-curve shows how Jan's present realisation of past experience, fades off with time.

Third we can impose a standard memory-curve on top of Jan's varied impressions (or information bits accumulated?) up to the midweek point. The memory curve is an intuitively obvious type of diagram, first proven by psychologist H. Ebbinghaus in 1882.[13] Here it depicts how Jan's weak perceptions (i.e. those of little interest) are soon forgotten as they fade away beneath the memory threshold. In contrast her stronger perceptions (i.e. of higher interest) persist far longer in her conscious memory.

Fourth – Though our build-up of this diagram has been totally conventional so far, Jan's frequent intuitions suggest that one crucial addition is still required. These pre-call or anti-memory experiences can then be expressed through a second, or future-oriented, anti-memory curve - one extended out to the right and more steeply up from Now.

In effect therefore Jan has now become a modern version of ancient Janus, the two-faced Roman God who could look both forward and backward into time! [14]

Naturally too this new paranormal curve must be much steeper, than its normal or past-oriented one. This is because Jan's intuitions are far less frequent, less detailed and much weaker, than their conventional past-oriented memory counterparts

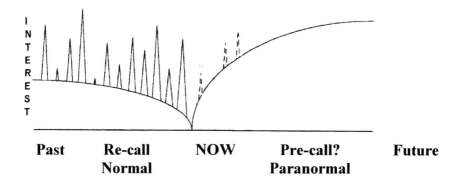

DIAG. 4-4: Intuitions suggest that a steeper, future-oriented, anti-memory curve should be appended, conferring new partial symmetry in time.

This final composite diagram also expresses two of those secondary facts of intuition I'd noted previously. First it depicts how 'slight' intuitions – i.e. those with mundane content of no great interest – tend to be complemented by reality very soon afterwards. For example:

> "I suddenly thought of distant Mary – and almost immediately she phoned!"

In contrast high-interest events (e.g. premonitions of drowning or a disaster like the Titanic) tend to be anticipated by weeks or months, as indeed other psi researchers have long realised. [16]

Secondly this diagram clarifies how full information about any experience is never really available to memory in either mode. Indeed with past incidents our normal memory, or re-call for detail, is at best very partial, as judges and accident police well know.

With future incidents our paranormal anti-memory, or pre-call capability, is liable to be even weaker. Which also of course, partly accounts for why we know so little about them still!

New estimation of frequency

And finally our diagram enables a first ever estimate, of the relative frequency of intuitions in everyday affairs. So just how relatively frequent seems paranormal *pre-call,* in contrast to the much more common or normal *re-call* capability?

To answer this question we need merely allow an average person like yourself about one act of re-call every second, and regardless of whether you're sleeping or awake. Further there are 86,400 seconds in a day, or 2,592,000 in a 30-day month. On average therefore we make something like 2.5 million acts of re-call within that period.

In addition, or as my first Coincidence Survey suggested, people on average may notice perhaps one incident of potential intuition (or anti-memory as we can now call it) within that same interval. However we can be conservative and lower this estimate to just 1 potential experience of paranormal pre-call, between every 4 months and 4 four years! There will also be about 10-to-100 million, acts of normal re-call within these times.

This estimate for the *relative frequency* of pre-call then gives a core value of 1 in 10-to-100 million, when compared with the more familiar re-call faculty. Taking the first figure only for simplicity, we can write it as 1/10,000,000 - or 10^{-7} in more scientific terms.

To the ordinary person this figure of 10^{-7} this may seem a very slight figure indeed. In fact it would probably seem so slight as to be hardly worth bothering about. But in science anomalies of this seeming slight order have often proved their worth, and turned out to be very important in the end.

For example when Marie Curie first isolated radium from pitchblende around 1897. she could only extract 100 milligrammes from half-a-ton of ore. This was a 2×10^{-7} anomaly or yield.

Likewise the Michelson-Morley experiment on light's velocity was searching for effects as small as 1 in 100 million (10^{-8}) – and failure to find them was one large factor in the general acceptance of Einstein's new Theory of Relativity (1905).

The lesson is therefore that anomalies as slight as 10^{-7}, had often opened up whole new scientific territories. So who could put limits on where the equally slight anomaly of intuitions might eventually lead?

A brief summary

All intuitions start to "make more sense" as a time-inverted version of past-oriented memory. This is a comparison with which some philosophers have toyed before. But a deeper analysis must reject that traditional term 'precognition', as misleading and confused.

Intuitions are better described in terms of anti-memory, or pre-call in verb form. For example the new term 'pre-call' permits a first ever estimation of relative intuition frequency. This works out around 10^{-7}, as compared with the normal re-call process.

New terms once properly defined often become a first step towards large open-ended progress, as for example when 'vector' and 'energy' were introduced into physics 150 years ago.[15] So too it proved with 'pre-call' or 'anti-memory', as I will now proceed to show…

PART TWO:

THE GREAT MISTAKES

"Many errors, of a truth, consist merely in the application of the wrong names of things"

Baruch Spinoza (1632-77)

5

CLAIRVOYANCE – THE MUDDLED INFERENCE

"When Victor is in a trance, I do not know anyone more profound, more sensible, more clairvoyant!"
Marquis Chastenet de Puysegur, 1784

EVEN THOUGH I HAD JUST discarded all of them with good reason, those notions of *'clairvoyance'*, *'telepathy'*, *'ESP'*, etc. had still known many believers down the years. But now they all so suddenly seemed quite misleading, whence had these words, and their associated false notions, come from originally? And why had they stayed so widespread for so long?

And above all would my new concept of anti-memory, which I had deduced from my purely personal analysis, still hold true for the greater world of all other intuitions elsewhere?

And so I began a long period of research, one stretching over too many years. It involved delving deeply into old studies of intuitions as carried out by others, analysing the various interpretations they proposed.

Eventually it then grew clear that there had been six main episodes of intuition research, spread over the last two centuries or more. Five of these episodes had promoted paranormal space transgression, in contradiction to what normal science holds. These had gone under different names like *'clairvoyance'*, *'telepathy'*, *'synchronicity'*, *'remote-viewing'*, *'ESP'*.

In fact just one other brief episode had ever considered pure time transgression as I had been led to do.

Over the next six chapters I'll therefore consider each of these investigative episodes in turn. It will then become clear that anti-memory was always the one single postulate required throughout. My own purely personal conclusions were therefore strongly reinforced to this degree.

People think "space before time"

But before commencing on this rather drastic revision of old research and conclusions, we need to consider one minor law of psychology and science history.

It's a law first noticed by a famed Irish exponent of Einstein's Relativity Theory. He was Dublin physicist J.L. Synge (1897-1995), the nephew of an even more famous Irish playwright. And in 1959 he wrote a seminal article in *New Scientist*. [1]

This was the start of the Space Race years. But, since we inhabit a Universe characterised by matter displayed in space and time, Synge wondered why time had always been so comparatively ignored.

For this avoidance or demotion of time by scientists, Synge blamed Euclid (300 BC) – the founder of textbook geometry which first described the laws of space. Most later physicists also start off from school imbued with Euclid – which may render them unconsciously biased towards space, while comparatively ignoring time!

Synge's article marked the start of a small flood of serious time books, of which Stephen Hawking's *Brief History of Time* (1987) is merely the most famous so far. Much deeper and more informative however was *The Natural Philosophy of Time* (1961) by London mathematician G.J. Whitrow. In this, and his subsequent *Time in History* (1988) he pondered a simple but little known rule of psychology and science history:

> People tend to start thinking in terms of space before considering time.

This may be because all human infants seem to arrive well equipped for comprehending spatial relations, while still showing very little appreciation of the temporal.[2] Ask any mother wakened up thrice nightly during her child's first year!

In any case nearly all later adult scientists have rather unthinkingly adopted the same priority. So that for example NASA now spends many billions on exploring space - but little or nothing on exploring time. This is indeed a most striking bias, one that future historians may well regard as by far the most curious omission of this scientific age.

Or as Whitrow summed up this whole affair. [3]

> "Men have strained their ingenuity to devise means whereby the peculiar characteristics of time are either distorted or ignored.
> "Thus ignoring the temporal aspects of nature...(they) have sought to explain it away in terms of the spatial, and in this they have been aided by philosophers"

All of which also could almost have been written, with most traditional paranormal research in mind. For, as I will now proceed to show, the same 'space-before-time' priority, has also pervaded two centuries of investigations there.

In addition to which all researchers have always been subject to that *Prime Assumption* which I've clarified in Chapter 4 – that Mind or memory can just reflect the past alone....

Swedenborg's vision

In any case the first scientific stab at intuitions can be traced to English polymath Francis Bacon who died in 1626. Now regarded as one of the founding fathers of modern science, Bacon took intuitions seriously. In fact he was first to suggest that this *'binding of thought'* might be proven by guessing cards - even if *'it hit for the most part, though not always'*. And indeed cards still form an excellent test for intuition, as we'll see in Part Three.

A century after Bacon, Emanuel Swedenborg (1688-1772) was a sort of Swedish Leonardo, who turned to mysticism in his later years. From which he was soon credited, with paranormal knowledge about far-off scenes.

Swedenborg's most famous such episode was focussed on the Great Fire of Stockholm - which took place on July 19, 1759. (Most North European cities were made of wood in those days, so that Great Fires were not uncommon in them all.) The tale was told in several versions by philosopher Immanuel Kant (1724-1804), not always the most accurate in such reports: [4]

> Returning from England, Swedenborg disembarked at Gothenburg in the afternoon. Mr.Castel invited him and some 15 others to his house.
> At about six in the evening Swedenborg absented himself, and came back a little later pale and aghast. He said that at that very moment a great fire was raging in Stockholm 300 miles away, and spreading rapidly.
> Around 8 o'clock he went out again and came back more joyfully: "Praise be to God the fire has been extinguished, and that three doors from my own house!" he exclaimed.
> Two days later a despatch rider arrived from Stockholm, with letters detailing the fire just as Swedenborg had said!

Apparently Kant thought that Swedenborg's mind was somehow transcending 300 miles of space or distance in this episode. In modern terms this would be somewhat as if one half of his mind was acting like a TV camera at the scene of the fire in Stockholm. And further reporting back to the other half, which was acting as a receiver in distant Gothenburg!

But of course it's much simpler to think that Swedenborg was merely pre-calling what news the messenger would bring two days afterwards. His vision can thus be subsumed into that common pattern I've deduced from personal analysis in Part One. So that my hypothesis – that anti-memory is the sole process required for intuitions of all sorts – therefore survives this new test of Swedenborg's vision quite unscathed.

And the same likely holds true for all records of supposed 'clairvoyance', as I'll now further show. Though we can't be quite fully certain in this latter case – because virtually all of the surviving records are crucially incomplete.....

Mesmerised mind-states

In any case by the time Swedenborg died in 1772, three natural forces were well established on the scientific scene. Gravity, magnetism and electricity all obviously transcended space, although by methods still not understood. This was the scientific climate in which one Friedrich Anton Mesmer (1734-1815) completed his studies at the University of Vienna.

For his final medical examination, 32-year-old Mesmer wrote a thesis advancing his own unique idea. He proposed that the gravitational attractions of the planets could also affect human biology, postulating a 'universal fluid' supposed to permeate all living things. A century earlier Newton had explained how gravity rules the motions of the planets; Mesmer was now proposing a similar process for people as well!

'Invisible fluids' were currently invoked for the newly emergent science of electricity. Gravity was also prominent because of Newton's recent clarification of planetary movements. So that Mesmer's proposed new 'universal fluid' was really a combination of these latest science concepts with older notions of astrology.

Soon Mesmer shifted from gravity to magnetism as the basis for his 'universal fluid'. This gave rise to the notion of "animal magnetism", apparently what we would now call hypnotism. For a few years after 1778 he flourished in Paris, so that investigations into 'mesmerism' thereafter became a predominantly French pursuit.

One interesting by-product was that the "mesmerised" state (for which read hypnotism) seemed further to confer some people with new mental powers. These became known as the "higher phenomena" – and were thought to include *'pre-vision, thought-reading, second-sight'*. For example:

> A certain society lady was sad because her pet dog had gone astray. But under the mesmeric influence she suddenly summoned her maid, instructing her how to find the animal.
>
> The maid was to go out in the street, summon the gendarme, and tell him to proceed to another street 15 minutes away. There he would meet a woman carrying a dog which he should then bring back with him.
>
> All happened as the lady had imagined, and the dog was duly identified as her own!

Once again of course there's no need to invoke *clairvoyance or telepathy* in this case. (That neither term had yet been coined in the modern sense is neither here nor there!) It's simpler to think that the owner was just *pre-calling,* how her dog would later be returned.

Clairvoyance shifts meaning

Mesmer himself seems never to have been greatly interested in such manifestations of the "higher phenomena". But, from about 1784, they were taken up with great enthusiasm, by one Chastenet de Puysegur, a country physician who managed a large family estate near Soissons in north-east France. There he grew interested in mesmerism for its healing powers. And he was lucky in quickly finding a subject most amenable to mesmerism! [5]

23-year-old Victor Race was a tenant shepherd on the family estate. Of simple mind and tongue-tied before his social superiors, he was *"quite the most limited man in the country"* as de Puysegur said.

However under the influence of mesmerism, Victor exhibited a startling change in personality. His normally placid countenance lit up with a new glow of intelligence. His shy, halting, speech grew rapidly more fluent. Victor's new mode of expression, his diction now flowing without any trace of stress or embarrassment, astonished all who knew him formerly. Or in short his whole demeanour was much changed.

Victor could also now apparently use 'pre-vision' to diagnose illness, and sometimes know songs which had just come into his master's mind. In short under mesmerism he seemed far more intelligent than when in his normal state. Or as de Puysegur wrote:

> "When Victor is in a trance, I do not know anyone more profound, more sensible, more clairvoyant" [6]

And with this single sentence a new and much muddled notion entered into popular usage, a change in semantics destined to influence countless millions ever since….

The context of these words suggests that de Puysegur was employing the term *clairvoyant* (Fr: *clear seeing*) in its original or traditional sense. The word applied to those who were especially wise or intelligent: they could *see clearly* into the nature of things much better than most people could. [7]

Great savants of the era - like Benjamin Franklin or Anton Lavoisier – would then rightfully he termed clear-seeing, i.e. clairvoyant. As indeed so too might Victor Race, when he began to show signs of enhanced social intelligence and skills.

After this however the word began to be associated with paranormal intuitions in various forms, those 'higher phenomena' sometimes manifested under mesmerism.

Another investigator of this era was Dr. Petetin, whose book on *Animal Electricity,* seems an obvious extension to de Puysegur's own work on *Animal Magnetism.*[8] Petetin reported on mesmerised people who could describe cards, medals, etc. placed out of sight. At times also their intuitions extended into other rooms, locked drawers, or even the pockets of visitors:

> And when the local doctor saw Madeleine in her mesmerised state, she was able to prove her ability by finding a gold watch, which he had hidden in a drawer.
> In addition she told him he had a bottle of white wine hidden in his greatcoat pocket, to which he assented in some wonderment.
> "And it comes from Chondrieu" she also said. This was a detail not then known to the doctor, but which he afterwards confirmed.

For such cases Petetin invoked a mysterious 'higher phenomenon of eyeless vision' operating over distance. He thought it could also pry into places where normal eyesight couldn't possibly operate. He never at all seems to have considered the alternative time option - that his mesmerised subjects might merely be anticipating, what they would soon learn about afterwards!

Which therfore affords yet another example, of that 'space-before-time' priority in common thought, as I've clarified above.

Historically in any case the semantics of 'clairvoyance' now grew confused. Originally the term had denoted unusual ability to see clearly into the nature of things, as Newton or Leonardo could. In such cases to 'see' was meant only in a metaphorical or intellectual sense.

But under mesmerism to 'see' was now interpreted more literally. It denoted a supposed ability to somehow 'see' into places where normal vision (either physical or intellectual) shouldn't be able to operate at all!

And yet in a third sense 'clairvoyance' is still a good descriptive term. This is because the required psi-state is characterised by unusual mental clarity, one freed as much as possible from all unwanted intrusions or mind-noise...

In any case all across 19th-century Europe people learned and otherwise began writing books attesting to the reality of those 'higher phenomena'. Unusual powers of vision often featured in these descriptions, this being the era when the notion of 'clairvoyance' in its modern connotation became widespread.

In English the term 'clairvoyant' became accepted by about 1850, introduced by writers like Lady Carlyle and Horace Walpole. In America it was brought in by Ralph Waldo Emerson as part of the Transcendalist movement.

Everywhere also despised fortune-tellers began reinventing themselves as more glamorous 'clairvoyants', the term still carrying those older connotations of clear thinking and intelligence. But semantically a crucial shift in accepted meaning had taken place with de Puysegur's report....

In science however, this muddled notion of clairvoyance went nowhere for the next 200 years. Then in 1971 its development was suddenly taken up in secret for Cold War spying by America's CIA. But now it was renamed 'remote-viewing' - a term more in keeping with the modern Closed Circuit TeleVision (CCTV) age. However, as we'll see in Chapter 9, 'remote-viewing' also came to nothing in the end.

In this it resembled the old muddled notion of clairvoyance which it replaced, being always a notion still fatally confused!...

Anti-memory throughout?
Reverting then finally to those first mesmerists, it's important to realise that they were mostly ordinary medical doctors, while few or none were scientists. So they would have known little of basic science principles, nor procedural aims like common patterns or minimum hypotheses.

It's therefore no surprise to find that the single most salient detail in their 'clairvoyant' episodes – i.e. later confirmation of the earlier intuition – was neither realised nor recorded by any of them. And because this single most crucial detail was never recorded, it's now quite impossible to state categorically that it always occurred.

Still we can infer from the circumstances that it probably happened, in most episodes at least. For example when Victor Race intuitively started singing a song his master was then contemplating, we can infer the latter's very natural exclamation of surprise.

(Though equally in this case it may have been de Puysegur who was exercising anti-memory - i.e. for a song that he was about to hear!)

Likewise in the case of Madeleine, her intuition about a gold watch hidden in a drawer, was probably confirmed immediately by excited revelation of the same. So too for her divination of a wine bottle hidden in the doctor's pocket, as also his confirmation of its origins.

Assuming my reconstructions of these 'clairvoyant' episodes are reliable, their common pattern is again what I clarified from my personal Surveys (Ch.1 - 2). Since all of them likely involved later confirmation of an earlier intuition, they're most simply considered as cases of pre-call or anti-memory,

In sum therefore those first mesmerists seem to have discovered, that intuitions can sometimes be maximised under hypnotism. But they quite failed to note the full facts of these intuitions as they took place. So that their conclusions were always fatally compromised. For they never considered the possibility of time violation as the primary process!

My previously personal conclusions therefore survive this first wider test of 'clairvoyance' quite unscathed....

A brief summary

Gravity and magnetism were two prominent topics in 18th-century physics, both of them obviously transcending space in some mysterious way. So that, around 1785, the first mesmerists (who were really hypnotists) naturally imagined they were doing something similar.

And when various hypnotised subjects proved to have heightened powers of intuition, some mysterious and space-transcending mental power was consequently invoked.

But all these 'clairvoyants' likely got later confirmation of their earlier intuition at some point. In which case anti-memory would always have been the sole and only process ever required...

6

TELEPATHY – A MYTH FROM TELEGRAPHY

"It is the most important work which is being done in the world...by far the most important!"
Prime Minister W.E. Gladstone – 1885

THE SECOND STRONG TEST for anti-memory is provided by the notion of *telepathy,* an idea that started in Britain during late Victorian times. There scientific interest grew in the parlour game of 'thought reading' - a notion popular since those 'higher phenomena' first manifested under mesmerism.

During these years also organised religion began to lose ground to the growing forces of materialism. And, as is still happening nowadays, some people considered that the new materialists were claiming too much and overstating their case. As for those 'higher phenomena' - with 'thought-reading' prominent among them – the materialists seemed too hasty in their dismissal of them all/

'Psychic' affairs

A few scientists also shared this unease. Their ranks included Oliver Lodge (1851-1940) who would later research early radio. There was also William Crookes (1832-1919), pioneer of the familiar vacuum tube still used in older TV sets and monitors. In their spare time these scientists started to investigate those 'higher phenomena' so long associated with mesmerism.

Crookes then reported positively on several people who could demonstrate intuition through card-guessing routines. He speculated that these abilities might emanate from a new kind of force: [1]

> "to which I have ventured to give the name of Psychic" (Gr: 'of mind or soul')

Crookes thought that such findings contradicted the canons of science, so that the latter would have to be drastically revised. But more cautious believers, like the great French Nobel physiologist Charles Richet, (1850-1935) disagreed: [2]

"(the new phenomena) are unusual, they are difficult to classify. But they do not demolish anything, of what has been built up so laboriously in our classic edifice (of science.)"

A sentiment incidentally, with which all my own conclusions and findings about anti-memory agree.

But in any case Crookes' new term 'psychic' quickly caught on. In Britain and then America it largely replaced the more continental notion, of 'higher phenomena' exhibited under mesmerism. Soon too it became invested with all those vaguely spooky connotations associated with it still.

And by then it was clear that 'thought-reading' and similar feats could occur, without any mesmeric or hypnotising process.

Another investigator of this era was William Barrett (1844-1925). He was a professor of physics who researched magnetism at The Royal College of Science, Dublin. As such he came to modify Mesmer's notion of 'animal magnetism' from the century before. Barrett now reasoned that 'thought reading' must involve 'thought transference' between two minds. As such there must also be paranormal transmission between the two.

Such reasoning must seem unduly mechanistic in our day. It derived very obviously from the electric telegraph (1837) and the telephone (1874). Both of these worked through sender-trans-mission-receiver components. And so paranormal transmission of thought was likewise considered to require a mental sender and a mental receiver. This was an obviously limited form of reasoning, although accepted by some leading parapsychologists still.

In all this however Barrett apparently never realised that his intuitive 'thought receiver' mostly - or more likely always? - attained later confirmation of the original idea. In which case his or her thoughts exhibited that usual common pattern I've established for intuitions of all sorts. They could be interpreted more simply in terms of anti-memory.

In any case this dubious notion of 'thought transference' was accepted almost immediately by the Society for Psychical Research (SPR), when it was formed in London, 1882. There too it was soon dignified by the new term 'telepathy'. The inferred similarity with telegraphy (or telephony) was thus formalised...

The SPR began with a group of Cambridge intellectuals – mostly learned in classics and philosophy – which was interested in psychic aff airs. It got off to a fine start during early days. Among its supporters were W. E. Gladstone and Lord Balfour who were Prime Ministers of the day.

On the more active side there was Frederick W. H. Myers (1843-1900) – a poet, classicist and later school inspector. He was the man who coined the term telepathy (lit: kindred feeling between distant minds).[3] For this term he remained an active proponent (some might even say propagandist!) all his life.

Around 1892 Myers also introduced 'precognition', more or less as an afterthought, to explain those few intuitions for which telepathy could not be inferred.[4] This term 'precognition' was borrowed from theology, though with little justification as we've already seen (p.35).

Intuitions surveyed

At an early stage the new SPR formed 5 committees – to investigate hypnotism, spiritualism, apparitions, sensitive abilities, thought transference. Apparently nobody was especially concerned with 'pre-vision' or 'premonition', the term 'precognition' not yet being introduced. Which again also of course just serves to illustrate, that common 'space-before-time' priority we've now met repeatedly.

A more productive act of the early SPR was to poll the public for anecdotes of intuitions or other strange phenomena. Soon 5,700 responses had been received, nearly all from the more educated classes to whom writing then came easily. It was therefore established for the first time, that intuitions seem to be quite common in everyday experience, and in addition that no mesmerism is required.

Often too, and as still today, these intuitions left those who experienced them very curious, sometimes almost desperate, to have them explained.

F.W. Myers, Edmund Gurney, and Frank Podmore undertook the very arduous labour involved in checking out many of these intuition anecdotes. They termed such incidents spontaneous, because they seemed to come from nowhere and quite unexpectedly. Then in 1886 they published some 700 anecdotes in their book *Phantasms of the Living*. This undoubtedly monumental work comprised two thick volumes, numbering at least half-a-million words in all.[5]

Phantasms was and remains the most comprehensive collection of intuition anecdotes ever published anywhere. And as such it provides an excellent reservoir for testing out the new anti-memory concept.

The general tenor of these anecdotes also conveys a strong sense of middle-class Britain in Victorian times. For example here's an abridged account from a young woman evidently accustomed to country pursuits:

> One Tuesday in the morning I received a letter from a friend, saying he was going to hunt that day.
> In the train I shut my eyes and presently the whole scene suddenly occurred before me – a hunting field and two men closing up to jump a low stone wall.
> My friend's horse rushed at it, could not clear it, and blundered onto his head, throwing off his rider, and the whole scene vanished. I was wide awake the whole time.
> On Thursday morning I received a letter from my friend, telling me he had had a fall, riding at a low stone wall, that the horse had not been able to clear it, and had blundered on to his head, that he was not much hurt, and had later on remounted.

Though this sequence was interpreted in terms of 'telepathy', it might have been termed a case of 'precognition' or even 'premonition' (forewarning) equally. More simply however it can now be understood in terms of anti-memory: the narrator was merely pre-calling the details of a letter she would receive two days afterwards!

Others examples of intuition were more trivial and even humorous:

> The Bishop's wife dreamt she had gone in for breakfast as usual after morning prayers. And then discovered an enormous pig, standing between the table and sideboard!
> So amused was she by this dream that she related it to the governess and children before prayers began.
> Afterwards she went into the dining room as usual – and there found the pig standing as in her dream. It had somehow escaped from its sty and wandered in there during prayers!

And finally *Phantasms* relates other intuitions of more familiar content. This is because they're still happening to ordinary people worldwide every day:

> Mr. Skirving was foreman over the masons working on Winchester Cathedral. One day he suddenly felt an intense urge to go home.
> He didn't really want to go because of the time and money he would lose. But eventually the urge became so strong he couldn't resist it any more.
> On reaching home he found that his wife had been knocked down, by a horse-drawn cab in a nearby street, was now lying injured and had been calling out his name. However she calmed down on his arrival, and later made a good recovery.

From a total of ca. 700 such anecdotes, the *Phantasms* editors cited 109 as a first proof for 'telepathy'. Nevertheless every last single one of these can be equally interpreted in terms of anti-memory or pre-call. For always the narrator receives later confirmation, of some earlier intuition, in some ordinary way.

Nor could this really have been otherwise. For unless all these narrators of intuition had received later confirmation of their earlier ideas, how could they have any paranormal tale to tell?

Nevertheless, like 'psychic' a dozen years earlier, Myers' new term 'telepathy' soon caught on. Its ready acceptance was probably conditioned, by those two other new terms – telegraphy and telephony – to which people were just getting used.

Indeed in America the celebrated author Mark Twain – who had many personal intuitions and once worked as a telegraph operator – used to refer to them as instances of 'mental telegraphy'.

Soon too the advent of wireless (1887) likely convinced even some doubters, that 'mental radio' was a very real possibility.

However this notion of 'telepathy' was always a vague and scientifically unjustifiable one. Partly this was because its proponents never realised that universal common pattern, of later confirmation always. Yet it was always observably present in every case of spontaneous intuition they had placed on record!

Once again we can see this quite easily, if we recast the 'pig-in-the-parlour' story to a less memorable end:

> The Bishop's wife dreamt she had gone in for breakfast as usual. And there found an enormous pig standing between the table and sideboard.
>
> So amused was she by this dream that she related it to the governess and children before prayers began.
>
> Nothing much unusual however happened on that day afterwards. She discovered no pig in the parlour, nor anything even remotely similar.
>
> And since she now had no strange intuition tale to tell, her silly dream was soon forgotten, by all concerned!

The essence of intuitions is therefore always foreknowledge, and always there must be confirmation at some future time. Conversely if such later confirmation is lacking – as in our revised pig-in-the-parlour anecdote above – there's simply no intuition tale to tell…

More 'telepathic' episodes

Telepathy seemed also supported by various positive card-guessing experiments reported in *Phantasms*, along with successful reproduction of drawings kept out of sight nearby. Their design was usually to have a 'sender', supposedly 'transmitting' to a 'receiver', who would then try to reproduce whatever information was involved.

But again of course, at least most - and more likely all? - of these experiments incorporated the same logical defect as before. That is to say the 'receiver' (who'd supposedly picked up information 'transmitted' by telepathy from the 'sender') was usually informed about the real nature of the target shortly afterwards. In which case he or she was then probably just pre-calling this later information once again.

However we can't now be fully certain about later confirmation in all of these early SPR experiments. For, as with 'clairvoyance' previously, this all-important factor is never commented on specifically. So presumably it was never really observed. Though again we may reasonably infer it, for most of these old reports.

'Telepathy' was also inferred for a long series of demonstrations, reported between 1910-1929 by the distinguished Oxford classicist, Professor Gilbert Murray (1866-1957). He'd learned how to demonstrate intuition, within a close circle of family and friends. This was a variation of the old Victorian 'thought- reading' game: [6]

> "The method was always the same. I was sent out of the drawing-room, either to the dining-room or to the end of the hall, the door, or doors, of course being shut.
> "The others remained in the drawing-room: someone chose a subject, which was hastily written down, word for word.
> "Then I was called in, and my words written down. A typical case record was Countess of Carlisle (agent): "Thinking of the Lusitania."
> G.M. " I have got this violently. I have got an awful impression of Naval disaster. I should think it was the torpedoing of the Lusitania."

Most of the other episodes concerned similar startling, i.e. high-interest, themes: Some 500 such were recorded over many years, and eventually published around the 1930s by the SPR. One-third of this total were considered wholly successful, one third partly successful, and one-third failures.

But again of course – and this time with no uncertainty whatsoever – Murray always received later confirmation soon after his intuition was declared. So that once more anti-memory was the one simplest option required throughout!

Another prominent investigator around 1940 was G.N.M. Tyrrell, an engineer who constructed the first 'guessing machine'. It was a series of 5 closed boxes, of which only one would display a light when opened.

This he tried out with his ward Miss Johnson, a young lady regarded with good reason as a strong natural intuitor. [7]

Miss Johnson then proved able to beat this machine consistently, and to an enormously improbable degree. It didn't matter whether her observer first knew which box was lit up ('telepathy?'), whether nobody knew ('clairvoyance?'), or even when the box would light up only after her choice was made ('precognition?'). She was equally adept throughout.

Considered in terms of anti-memory however, the key point is that apparently always she received later confirmation of her earlier intuition - by open all boxes after 'guessing' to find the lighted one. So that once again anti-memory survives this test, which can be taken as further proof of its reality.

Telephone telepathy
The best attempt at demonstrating telepathy in recent times has been a series of telephone experiments conducted by Rupert Sheldrake, by far the most prominent British researcher into psychic matters today. He seized on the fact, that one of the most common of all intuitions, is knowing who is about to telephone! As for example in my own first anecdote reported on Page One.

Sheldrake, with co-worker Pamela Smart, proceeded to quantify this effect in an impressive series of numerate experiments.[8] A subject at a telephone was asked to 'guess' which one of four persons was about to call, the latter being chosen randomly. Though of course his or her identity was then immediately obvious, once the intuitor had picked up the phone!

Where 1-in-4 scoring (25%) would then have been expected by sheer chance, 53% was the highly significant average when both participants knew each other well. Further this competence level held true over great distance: it worked even when people phoning were several thousand miles away!

Rupert Sheldrake prefers to interpret his findings in terms of 'telepathy' - while still acknowledging the ever-present 'precognition' option. Still it appears to me that the latter option must seem preferable, primarily because it seems more logical.

This is because the distance interval varied greatly but the strength of the effect did not. It was just as strong at 1,000 miles compared with 10!

On the other hand the time interval – between 'guessing' and lifting the phone to hear who was really at the other end – remained largely constant at ca. 10 seconds throughout. As such time interval correlates very well with that constancy of scoring, whereas space interval just doesn't correlate at all. [9]

The clear conclusion must therefore be that time violation, rather than space violation, was the one paranormal process at work throughout. Which again makes anti-memory – or pre-call of whom was going to telephone next – the most simple or parsimonious interpretation required.

An elusive state of mind

Mistaken of interpretation though it may be, the vast literature of telepathy is still valuable in a different way. It reports a good deal of information and impressions, on the background mind-state from which these intuitions came. For example here is Gilbert Murray describing how he did his feats:

> "The conditions which suited me best were in many ways much the same as those which professional mediums have sometimes insisted upon....
>
> "I liked the general atmosphere to be friendly and familiar; any feeling of ill-temper or hostility was apt to spoil an experiment. Noises or interruptions had a bad effect...
>
> "(My impressions) always begin with a vague emotional quality or atmosphere. That is, it was not so much an act of cognition, or a piece of information that was transferred to me, but rather a feeling or emotion.
>
> "And it is notable that I have never had any success in guessing mere cards or numbers, or any subject that was not in some way interesting or amusing."

Very similar observations were made, by American self-taught intuitor Mary Craig.[10] She taught herself to divine and then draw pictures, being "sent telepathically" to her from up to 40 miles away. Though yet again of course, since she always viewed the originals after attempting to reproduce them, anti-memory is a more simple inference.

In addition her reproductions of the pictures were never correct in total detail, which substantiates the parallel with normal memory yet again!

Nevertheless Mary Craig's observations on how she did it are both penetrating and acute. They're also remarkably similar to what later 'remote-viewers' would report (Ch.9). And as such also they're highly relevant, to similar experiments with controlled intuition I'll be reporting in Part Three:

"You must be able to enter a state of complete relaxation coupled with undivided concentration on the unknown target you're trying to divine....inhibit all thoughts from outside and just relax by 'letting go'.

"At first some practice is required to combine these two rather contradictory states. Then make your conscious mind a blank and watch what starts to form in it.

"Beware the tendency of your imagination to fill out the picture and make full sense of it too rapidly. There seems to be contradiction between spontaneous relaxation and concentration all at once.

"Perhaps we each have several minds, and one of these can be blank or asleep while the other controls it, then presenting some thought or picture agreed in advance, as it comes back to consciousness.

"It is possible to be conscious and unconscious at the same time!"

Somewhat similar observations, were reported by New York intuitor Harold Sherman, about his experiments in the 1930s: [11]

"I was able to bring to each sitting a natural enthusiasm and eagerness. This attitude seemed to create in me an energy I needed to get results...

(You must) make your mind as clear and passive as a pool of water so that it will reflect the vaguest shadow, or so it will react to the tiniest pebble which may cause a ripple on it."

This statement by Sherman communicates those Childlike aspects of eagerness, enthusiasm, energy - together with clarity and passivity - that are the main operational factors essential to success. So that - whatever it might mean in reality - it conveys as much of the intuitive mind-state, as any description that I've ever seen.

These three self-observant intuitors therefore support that other conclusion I'd reached in my own case - that intuitions tend to maximise in a rather special state of mind (Ch.2). It's hard to define because we lack adequate words for it. It's easy to recognise for those with some minimum experience. Its overall outlines are generally agreed.

Nevertheless all these other observers also consistently favoured telepathy as the process involved. But this may just provide yet another illustration, of that common space-before-time priority in most thought.

Once again however anti-memory is the much simpler concept required throughout. So that again it survives this second test of 'telepathy' quite intact: there's simply no logical need for the older traditional idea!

A brief summary

The Victorian notion of 'telepathy' seems an obvious derivation from telegraphy: it proposed purely mental components of 'sender-transmission-receiver' corresponding to the physical device. Still, in all those relevant anecdotes which were then collected, nobody ever highlighted the one clear common factor throughout. This is merely that all narrators always had earlier intuition, confirmed later by ordinary means.

All reports of supposed 'telepathy' in the literature, starting with 109 anecdotes cited by the early SPR, can therefore be interpreted more economically in anti-memory or pre-call terms. And at least most experiments can be explained likewise.

Proven intuitors further agree with my own first conclusions about the process. A rather special mind-state, which I've earlier termed the psi-state, is always necessarily involved.

7

ESP – THE BARREN OXYMORON

"Dr. Rhine's conclusions represent either one of the greatest mistakes of scientific history, or the most important piece of research done this century…."
New York Herald Tribune, Oct.3 1937

A THIRD TEST for anti-memory is provided by the common American notion of Extra-Sensory Perception or 'ESP'. Its origins go back to half-a-century after the SPR was founded in 1882. This was a time when Scottish professor William McDougall was appointed Head of Psychology, at the newly endowed Duke University, North Carolina. There he was able to establish the first ever university department wherein researchers could study paranormal affairs.

So that, after 1927, there began the third or American phase of intuition research.

McDougall hired a 32-year-old botanist called J. B. Rhine (1895-980) to help him in this task. The latter began by assuming that clairvoyance and telepathy were real and valid phenomena. Then he started numerous card-guessing trials with students, their sum apparently indicating these paranormal faculties at work.

Card 'guessing' like this was by no means unusual before Rhine's era. For example *Phantasms* reported card-guessing trials, which appeared to support telepathy, from as early as 1866. Likewise French physiologist Charles Richet reported some 3,000 trials – perhaps 50 hours work - with ordinary playing cards. He concluded that ordinary people could sometimes guess suits correctly, and to a significant degree.

But Rhine brought more American attitudes of mass production to this task. He employed dozens of ordinary Duke students in his card experiments. This showed that many ordinary people could 'guess' cards to a non-chance degree. But unfortunately their success was mainly slight or marginal.

So a hunt began for star performers, and over a few years in the early 1930s about eight were found. Three in particular - A.J. Linzmayer, Hubert Pearce, C.E. Stuart – proved especially consistent with high scores.

In addition Rhine replaced the old term 'psychical research' with the seemingly more scientific term 'parapsychology'. This meant the psychological study of mental events beyond current capacity to explain. It conveyed the rather doubtful premise that formal psychology was now the best way, to make more sense of psychic affairs.

The genesis of 'ESP'

Initially Rhine regarded the evidence as proving *telepathy* and *clairvoyance* more or less equally. The former seemed established from those earlier card-guessing experiments that had preceded him. The latter then seemed verified from his own later card-guessing results. So were his guessing volunteers somehow using *clairvoyance* or 'distant vision', to 'see' the card in the experimenter's hand?

To accommodate both possibilities Rhine therefore devised a new term - *Extra-Sensory Perception* or ESP. But he never seems to have realised the degree of oxymoron (i.e. self-contradiction) inherent in this term. For since perception means 'attainment of information through the senses', how possibly could it ever be extra-sensory?

In reality too the kind of information involved with intuitions doesn't really have the characteristics of a perception at all. Instead *"it emerges like a memory from the subconscious mind"*- as parapsychologist Rene Sudre noted after long research.[1] (A remark which incidentally supports the new anti-memory idea.)

In any event, the case for ESP was put forward by Rhine in his first book, *Extra-Sensory Perception* (1934). A more popularised version – *Hidden Frontiers of the Mind* – appeared in 1937, to be greeted with much public enthusiasm but strong scientific dispute. So started a widespread American interest in ESP, an interest that has only started to wane in recent times.

By 1937 however Rhine had further come round to realising that *'precognition'* also, could have been responsible for success in his card experiments. For by now he had found some students who could apparently 'guess' the next card *before* it became known to anyone. So might not *all* of his earlier successes have been due to guessers employing foreknowledge of the later outcome?

(Which is of course also precisely, what the new anti-memory outlook would conclude!)

But by now Rhine had committed himself to telepathy, clairvoyance, and his all-inclusive paradigm of ESP. So he never seems to have grasped the much simpler truth - that foreknowledge *only* might be the sole process logically required. Instead, in a classic case of compounding confusion into complexity, Rhine now added a third ESP function, to his first acceptance of telepathy and later addition of clairvoyance: [2]

> "The experimental confirmation...that distance was no limiting factor in ESP led to the inference that the ability should not be expected to be related to time either...independence of space would have to mean comparable freedom from time."

It does indeed seem rather difficult, to follow Rhine's reasoning here. But in any case once again this statement illustrates that common habit of thinking in terms of 'space-before-time', a trend we've now met several times. Further it violates that primary rule of Ockham's Razor (Ch. 3) – i.e. that *"mysteries must not be multiplied beyond necessity."*

Rhine's card routines

From which it also follows that Rhine's card-guessing procedures now need a new and more informed scrutiny. First therefore he sought to transform those few earlier and rather amateurish card experiments into something more professional. So he replaced their traditional playing cards – with all their numbers, suits and colours – by cards of a more neutral design.

Rhine's new cards were designed by one Dr. Zener, a psychologist who gave his name to them. He proposed 5 different and very basic black symbols – star, wave, cross, square, circle. These symbols were meant to be as neutral as could be.

Arguably however this was another large mistake. For neutral designs are also bound to be very boring - made purposely devoid of any interest or emotional content. Yet these are two of the clearest characteristics of intuitions, in their natural or everyday milieu.

From which one might argue that Dr.Zener was unwittingly minimizing, that very effect which he was trying to maximise!

In any case the guessing student had a 1-in-5 probability, of getting any particular Zener card correct. But - at first - Rhine found many students who could consistently outguess this average. For good performers the average success rate was something like 7 in 25.

Allowing for 5 chance successes on average, this meant they were using some other intuitive means to score an additional 2 times in 25. Or in other words intuition may have been operating about 8% of the time, i.e. around 1 'guess' in 12.

But on that important point of later confirmation – which all intuitions display in ordinary experience – Rhine's five books are almost totally silent. Not recognising this aspect as most crucial, he never really dwelt on it at all. So his books do little to clarify when, if indeed ever, the guesser became informed about the real nature of the target card.

However it does seem that at first, during those early and most successful days, the student 'receiver' merely spoke his guess. Whereupon the experimenter, or 'transmitter', recorded this choice - and then informed the 'guesser' of the actual result.

At any rate we can be sure this was the procedure in one of Rhine's most famous and successful trials. Here star scorer A.J. Linzmayer correctly guessed 21 out of 25 cards, while seated as a passenger in Rhine's car.[3]

> "Linzmayer called the card about 2 seconds after I laid it down, and I said 'Right' or 'Wrong' as appropriate"

Immediate confirmation is also apparent in an early procedure known as 'blind matching'. Here the 'guesser' tried to match up 5 cards in his or her own chosen order, with 5 others cards laid out in a row by the experimenter, but hidden behind a screen. Once this choice had been completed however, the reality was then revealed. As indeed a rather delightful photograph in one of Rhine's books still reveals. (*Parapsychology* – Blackwell – p.148)

When all of Rhine's early card-guessing reports are then totalled, the odds against chance are far beyond 1 billion, billion to 1. In fact the odds are so high that there's no chance, that chance alone could have been responsible! For if every last member of the human race were to repeat the exercise, it would still be many thousands of years before chance could afford anyone a similar result.

A point on which the American Institute of Mathematical Statistics agreed: [4]

> "Assuming the experiments have been properly performed, the statistical analysis is essentially valid. If the Rhine investigation is to be fairly attacked, it must be on other than mathematical grounds".

'Down-Through' Procedures

With the Linzmayer car episode also, as apparently in most early trials, both guesser and experimenter checked up to agree on the running total after every 5 cards had been drawn.

Soon afterwards however, possibly on spurious grounds of efficiency, this check-up period was extended to take in 25 cards at a time. The procedure then was for the guesser to work down through 25 trials before he was ever informed, about each singular result. This was the general basis of the 'Down-through' (D/T) procedure, which then became standard in all Rhine's experiments.

With early D/T. procedures however, a curious new phenomenon called *salience* was soon observed.[5]

> "One of the most normal aspects is salience – the tendency of the ends of a column or sequence to stand out in rates of success."

What this means was that successful guessers, say with a typical score of 7/25, tended to score best in the first and last segment of 5. Or in other words their extra, or non-chance, increment of 2 cards might appear at an average rate of about 1 each in these first and last segments.

But again of course this is just what one might expect, if it was really anti-memory that was operating throughout. Suppose for example that you were to try and memorise a long list of 25 cards in the normal or past-oriented way. Then it's those 'landmarks' near the start and end of the sequence, which you might expect to reproduce most accurately.

Anti-memory was also the most likely interpretation when star guessing student C.E. Stuart worked on his own through 7,500 trials with large success. In this episode immediate later confirmation must necessarily have taken place throughout.

Otherwise Stuart felt that his success was strongly dependent on mental attitude and self-confidence. This substantiates what we've already seen reported several times. Intuition stems from an especially delicate, labile and sometimes elusive, attitude of mind.

Altogether Rhine wrote five books on ESP between 1934 and 1960. None of these books present either facts or conclusions, nor indeed experimental details, in any coherent or easily assimilated form. But their sum is that nearly all successful card-guessing was conducted during Rhine's first decade in the 1930s, those early and most enthusiastic years!

Critics of no practical experience have often ascribed this decline in Rhine's results, to better and tighter controls. But to those more acquainted with intuition's many subtleties, it may seem more likely that decreasing enthusiasm was at fault.

For - as distinct from his later and more paranoiac years when he displayed strong signs of what I will term "intellectual arthritis" - the younger Rhine was a source of great enthusiasm. And as such he was likely to maximise in his volunteers, those Child-like attitudes whence intuition emanates.

This too is a most important matter for eliciting psi function, and as such one to which I will return later in Part Three.

In any case also, Rhine's striking decline in card-guessing results, may further correlate with another common factor in his experiments. This was his increasing, though hardly logical, insistence on 'down-through' (D/T) procedures after the first few years. D/T guessing involved guessing *down through* 25 cards at a time, but without any immediate or individual confirmation, as I've explained above.

And as this peculiar practice became standard, all guessing success seems to have died more or less totally away!

D/T guessing ability might always have seemed quite sensible to anyone who believed in ESP. In contrast its overall failure is only to be expected from the anti-memory viewpoint. To a slight extent therefore it might serve as a decisive test between the two paradigms.

But to fully substantiate my inferred correlation - between Rhine's later failure and his adoption of D/T protocols - would require long tedious working through all his many research articles. And this is a task I will gladly leave to others, with no qualms whatsoever because of the seemingly nonsensical nature of those D/T protocols....

Here too however one must note that not all of Rhine's reports are so readily interpreted in anti-memory terms. For example there was the long series conducted by experimenter J.G. Pratt with star guesser Hubert Pearce. In this series the general procedure was that Pratt turned over and stared at a card, while Pearce wrote down his 'guesses' in another building many yards away. Then they met and checked up on this D/T scoring after every 50 trials.

The overall scores in this trial were certainly significant, with an average of 30.22% (instead of an expected 20%) over 1850 trials. But to what extent salience could have been responsible is unclear. If it wasn't, perhaps Rhine's people were dealing with some higher order of mystery at this point.

Still this needn't trouble the anti-memory hypothesis unduly - because new paradigms are seldom adequate to explain all the facts at first. It's enough if they're able to deal with most of them!

Finally not all was completely negative in Rhine's card experiments. One worthwhile effect he discovered was the proven reality of involuntary 'missing' - i.e. sequences where guessers scored less than average to a significant degree. 'Missing' was and remains a most valuable concept. It's also relevant to controlled anti-memory in practice, as I will show later in Part Three.

Intuitions analysed

Rhine's near-total emphasis on card guessing was also perhaps a bit unwise. It meant that apparently he never gave much or any consideration, to clarifying those basic facts of intuition as manifest in the everyday milieu. So that for important factors like content, relative frequency, mind states, potential common patterns – none seem to have been really explored by him at all.

In effect therefore Rhine apparently never thought to check up, on those natural premises from which his future speculations would derive. So all his work is characterised by a strange lack of concern with natural intuition facts. Rather they were fixated on card guessing as the ultimate route to truth. He "aimed to bring (intuition) into the laboratory" as he often said.

Still this was all very much as if Rhine the botanist were trying to cultivate some rare and delicate wild bloom in his laboratory. But without first clarifying those natural conditions in which it flourished best outside!

In any case hundreds of intuition accounts soon started flowing into Rhine's laboratory, mostly from ordinary people sometimes almost desperate to have them explained. The result was a new collection of everyday intuitions, in number comparable to those collected by the early SPR.

As might be predicted, nearly all of these new reports were also very similar to those earlier *Phantasms* anecdotes: [6]

> While an American minister and his wife were travelling on vacation through Switzerland, she had a sudden unaccountable impression that her sister in Chicago was dead. But she didn't tell her husband immediately.
>
> A few days later came the equally strong conviction that her sister was being buried – and this time she told her husband.
>
> When the news from home finally caught up with them, she was right on both accounts...

Rhine's wife, Louisa, took charge of the increasing evidence from anecdotes of this kind. And after 30 years she published 186 of the more impressive accounts: [7]

> A housewife engaged in washing up took off her rings, and laid them on a shelf near the sink as usual.
> Later she went to look for them and found them gone. Helped by her husband she searched the garbage and all likely places to no avail.
> Then she had 'a funny sensation' in which something told her to look in the ice-cube tray. And there were the missing rings, frozen into an ice-cube!
> Earlier her husband had filled the ice-tray, in the dark with water from the sink, using a glass in which presumably her rings had been!

A rather different story came from a collector of rare minerals:

> Geodes are hollow stones partly filled with large crystals which have formed inside. One night the narrator, an amateur geologist, dreamt of an especially beautiful sample he could find.
> It seemed to be lying in shallow water, near a long gravel bar, in an unfamiliar river 15 miles away.
> Next Sunday he drove with his wife out to that district for a picnic, and after several enquiries found the gravel bank of his dream.
> There too he found his big beautiful geode, lying exactly as he'd dreamt of it the week before!

Finally Louisa Rhine reports another anecdote, where foreknowledge is the only real paranormal possibility:

> A young woman in the country dreamt she saw a plane crash nearby. In her dream it killed the pilot and set on fire a cottage which she knew. Also the dream fire engine from town took the wrong road and arrived too late.
> That evening she heard a plane overhead and sensed it was the one about to crash. She begged her husband to ring and warn the firemen which road to take – a request he refused.
> The plane duly crashed on the cottage and killed the pilot. And the fire engine duly took the wrong road as it raced to put out the fire.

All these incidents above the author interpreted in terms of various hypothetical kinds of ESP. Still they can all be understood more simply in terms of anti-memory. Further they can all be accommodated more specifically on the pre-call diagram (Ch.4).

The same applies to 179 - or 96% - of those 186 intuition anecdotes which Louisa Rhine relates in her book. In which case again one must be informed by Ockham's Principle, so that anti-memory is the single hypothesis more parsimoniously required.

As for the remaining 7 of Louisa Rhine's anecdotes which can't be interpreted in this way, these may represent some genuine residue of chance coincidence at work. Or again they might hint at mysteries of a higher order, ones totally beyond our comprehension at this stage. In science it's seldom that new paradigms can explain all the facts initially.

New emphasis on Mind

By the late 1960s in any case, many investigators were starting to abandon the oxymoron ESP, in favour of the more neutral term psi. Though ESP is still used by those less rigorous researchers even nowadays! [8]

Psi was a term suggested by Cambridge psychologist Robert Thouless, as I've stated before.[9] It's a noun derived from the adjective 'psychic', as introduced by SPR researcher William Crookes in 1871. It involved no preconceptions whatsoever, about what the real nature of intuition might be.

The late sixties also brought dawning realisation that Rhine had perhaps adopted the sterile routines of behaviourist psychology a bit too slavishly. There was a growing suspicion that perhaps behaviourist routines weren't really the best way, of investigating the super-subtleties of psi.

At last therefore researchers now finally started to take heed, of what accomplished intuitors like Mary Craig (Ch.6) had been emphasising vainly for too many years. Mostly these intuitors further agreed that endless guessing with Zener cards - which had been purposely been made so very boring – wasn't really much suited to the labile and delicate intuition faculty.

In 1974 this was the background, against which researcher Charles Honnorton aimed to maximise intuition, with new Ganzfeld guessing routines. Ganzfeld (Ger: uniform field) is a standard technique of deliberate information deprivation. Those subject to it may then enter a relaxed but fully conscious 'internal attention state'.

Or in other words they're encouraged to engage in deep phenomenology. This is much as I've described aiming for, but without any outside assistance, in Chapter Two.

For example a guessing volunteer may lie relaxed on a couch hearing white noise of no special pattern through earphones, while simultaneously wearing goggles through which red light is diffused. Often these goggles are split ping-pong balls, which seem to have been chosen as a telegenic factor for the media!

Like the traditional crystal ball used by clairvoyants, Ganzfeld now uses technology, to attempt what intuitors like Gilbert Murray and Mary Craig had attained unaided down the years. For intuition to manifest under deliberate control, you must try to "still down the rippled surface of the mind" as far as possible!

Given its new physical components and associated computer gadgetry, it's then disappointing to find that the psychological component of Ganzfeld testing still reflects very old routines. For, as first reported in Phantasms back in 1886, volunteer 'receivers' still mainly try to gain details, of some random picture being 'transmitted by a sender' from another room!

With Ganzfeld the judging routines, which decide success or failure, are also very cumbersome. So it takes a startling 4.5 man-hours to process just one single trial from the guessing volunteer. Small wonder then that, over the 30 years since it was introduced, there have been less than 6,000 individual trials in total. This works out at just 4 individual 'guesses' weekly, across the entire world community of 6 billion people overall!

Nevertheless, as researcher Dean Radin has summarised, Ganzfeld experiments have the best record for all 'guessing' experiments to date.[10] Over 24 years up to 1997, there were 40 publications reporting a total of 2945 individual trials. In sum they scored 33.2% success where chance expectation would have been just 25%.

For about 1 in 9 individual trials therefore, we can conclude that intuition was at work. Overall too the results give astronomical odds in favour of something significant going on.

More to the point in any case, that vital factor of future confirmation is again always present in these Ganzfeld routines. So much is made clear in passing, by Caroline Watts reporting from the Koestler Parapsychology Unit at Edinburgh:[11]

> "At the end of the sending period, the receiver looks at the target and three decoy targets, *not yet knowing* the identity of the actual target" (Italics mine)

Which also means that the guessing volunteer is informed about the true nature of the target soon after the trial is done! So that with successful Ganzfeld experiments, and despite all the new technology, it's simplest to conclude that anti-memory or pre-call has been operating much as usual.

After almost a century therefore, parapsychology remains very much a fringe subject on far science frontiers. It has never at all progressed at the same rate as, say electronic science. But the latter was at much the same proto-stage of development, when the SPR was founded back in 1882.

Meantime the simpler concept of anti-memory can suffice for at least most of those findings generated by the notion of ESP. The former is therefore again validated to this degree. Though whether or not those strange D/T routines are equally amenable must be left to others to decide.

A brief summary

Most of Rhine's research into intuition can be faulted on several grounds. First he never much studied intuition in its natural or everyday milieu. Instead he accepted 'telepathy' and 'clairvoyance', as proven entities from the start. Soon afterwards he further accepted 'precognition', lumping all three terms together in that peculiar oxymoron of 'ESP'.

This notion also made it quite natural to think that paranormal awareness could function, without any need for confirmation at some later time. Hence came Rhine's introduction of D/T or 'no answer' guessing routines. Upon which also his earlier card-guessing success largely died away.

Ganzfeld routines introduced later were more successful, and probably for two reasons. First they paid more heed to what natural intuitors had been saying all along. Second they provide later confirmation to their subjects, as indeed anti-memory would always require.

8

SYNCHRONICITY – A NEEDLESS MYSTICISM

> *"It is painful to watch how a great mind, trying to disentangle himself from the causal chains of materialistic science, gets entangled in its own verbiage."*
> Arthur Koestler - *The Roots of Coincidence* – 1972

Physics intervenes

A fourth test for anti-memory is provided by another paranormal notion, i.e. *synchronicity*. This non-experimental interpretation of intuition was proposed in the 1930s by Freud's former disciple, Swiss psychoanalyst Carl Gustav Jung (1875-1961). He was a follower of Kant and Swedenborg, as recent work has shown in large detail. [1]

Jung also lectured beside future Nobel physicist Wolfgang Pauli (1900-58), at the Technische Hochschule in Zurich. So a collaboration between physicist and psychologist emerged conveniently. The details of this collaboration are well told by Arthur Koestler in *The Roots of Coincidence* (1972), and I've based most of what follows on his work.

A few years earlier the physicist had proposed the well-known Pauli Principle. This proved to be fundamental to quantum theory, which describes the behaviour of electrons. But in the 1930s Pauli was troubled by the metaphysical, almost mystical, ontology apparently required by the new quantum theory which he'd just help devise.

And since mathematics and physics theories are ultimately creations of the human mind, Pauli wanted to study that connection more than most.

What concerned him mainly was the new quantum *acausal principle*. This held that events which involve singular electrons seem to happen without any obvious cause. In practice it would mean that no physicist, nor any law of physics, could ever possibly predict what any singular electron might do next.

Pauli suspected that this same quantum principle might also be extended into the larger world of everyday experience: [2]

> "It would be most satisfactory of all, if physics and psyche (i.e. matter and mind) could be seen as complementary aspects of the same reality."

In this he was a first seeker after some new psychophysical viewpoint, a new isomorph (Lit: similar shape) that might unite psychology and physics more effectively than before.[3] It's also one which anti-memory can now better provide. (Cf: Ch.19).

As part of this quest at an earlier stage, Pauli had written a study on mediaeval science and mysticism. His subject was Johannes Kepler (1571-1630), a famous astronomer who was also an astrologer. This study was then published by Jung's nearby Institute.

In seeking his new union between psychology and physics, Pauli veered off the conventional path of science to speculate on psychic happenings of all sorts. Poltergeists, ghosts, intuitions - all seemed to him to possess a sporadic and unpredictable nature, suggesting a link with quantum unpredictability.

Maybe they were just manifestations of acausality, on the much larger scale of the everyday world?

Throughout his lifetime Pauli was also apparently subject to 'psychic' manifestations of various sorts. For example his students joked about the 'Pauli Effect' – the tendency of their electrical experiments to go haywire whenever he walked by. It's been said too that all his life he was troubled by the significance of the figure 47 – and that this turned out to be the number of the hospital room in which he died! If true, this may have been an extreme example, of anti-memory at work?[4]

Jung's anecdotes

In any case Jung had complementary views. Like French philosopher Henri Bergson (1859-1941) he regarded intuition as a faculty not contrary to intellect, but rather supplementing it. And throughout his life he recorded those cases of coincidence or potential intuition which came his way, indeed somewhat as I've described my own two surveys in Part One.

Typical of these Jungian anecdotes was the following:[5]

> A young woman patient told Jung of a dream she'd had the previous night, wherein she saw three tigers seated threateningly before her.
> Jung proceeded to interpret this dream in terms of her strong power complex and "devouring attitude" towards people.
> Instead of which she actually did encounter three tigers that afternoon in a barn along Lake Zurich - a most unusual spectacle for the placid Swiss countryside. They were caged up as part of a travelling circus which had just come into the vicinity!

Once again of course this incident fits well into the general anti-memory concept, and also the more specific pre-call diagram. Most simply the patient was just dreaming, about what she would encounter a few hours afterwards!

Jung however never emphasised simplicity in his approach. So he lumped intuitions in with a ragbag of other occult phenomena quite indiscriminately. Prophetic dreams, déjà vu, foreknowledge, ESP, psychokinesis, horoscopes, oracles, omens, numerology - all were accepted equally by him!

Still Jung's collection of psychic anecdotes now complemented Pauli's interest in the same. And eventually this symbiosis produced Jung's treatise on Synchronicity: An Acausal Connecting Principle. A first version was published by Jung's Institute in 1939. Years later it appeared in book form in German (1952), followed by an English version (Routledge – 1955).

Though still much quoted by mystics and New Age gurus, Jung's book nevertheless proved a great disappointment overall. Its content is very much obscure, rambling, sometimes incoherent, hardly capable of further development. So that the result of their labours brought forth "a mouse instead of a mountain!" as Koestler wryly observes.

Synchronicity was in any case a very curious new word. Apparently it was first meant to describe the property of distant events occurring together at the exact same time, i.e. being synchronised. This was how Jung's book first defined it as -:

> "the simultaneous occurrence of a certain psychic state, with one or more external events, which appear as meaningfully parallel".

Their meaningful aspect is that they may coincide in time and have similar content, while still unrelated causally by any physical process.

There was however one immediate problem here. As the first SPR researchers had noted, most intuitions are seldom really synchronised with those external events to which they relate. They seldom happen together or simultaneously. For example, in the anecdote I've just quoted, above there was a time-lag of several hours, between the woman's earlier dream, and her later encounter with those 3 tigers.

Another anecdote, which Jung relates in his book (p.38) revealed a similar lack of simultaneity:

> An acquaintance in Europe experienced a very strong dream premonition, "around the same time" as a friend died unexpectedly while on a visit to America. Next day the premonition was confirmed, by a telegram which told of the fatality.
> Later however Jung investigated the exact timings of this incident. He found that the death was some hours before the premonitions came.

> In which they could hardly been intuitions of the death in actuality, but rather of the telegram which brought the news!

So Jung was now forced to conclude that most intuitions are "evidently not *synchronous* but rather *synchronistic*" In English at any rate, this may be taken to mean events *not exactly simultaneous,* but rather occurring *around the same time.*

To get round this evident lack of simultaneity, Jung next made a very extravagant or far-out proposal: our unconscious minds may function quite outside the ordinary physical framework of space-time! His oft-quoted tale of The Golden Scarab (p. 31) seemed to him somehow evidence for this postulate:

> Another young woman patient dreamt she was given a golden scarab, a version of the lowly dung beetle which the ancient Egyptians rather strangely adored.
> And as she related the details to Jung in a darkened room, there came an insistent tapping on the window from an insect hovering outside.
> He opened the window and caught the insect as it flew in to the light. It was a common rose-chafer (Cetonia Aurata) or scarabeid beetle, "the nearest analogy to a golden scarab that one finds in our latitudes"

Here again of course one can interpret this incident in those now familiar terms of anti-memory. In fact it fits the pre-call diagram perfectly. It then becomes merely another future-oriented dream intuition, with imperfect details of later personal experience in that darkened room. Nor is any further mysticism or cosmic significance required.

Jung also postulated that intuitions are most likely to occur in "crisis" situations like love, conflict, danger, death. All these of course can now be understood more simply, as merely high-interest events most likely to project in either past or future memory mode.

In addition a more thorough acquaintance with intuitions – or as I've reported in our first two chapters – would probably soon show that they're mostly concerned with everyday, trivial, mundane, concerns. Comparatively few intuitions then really reflect those big issues like death or accident, on which popular imagination tends to seize. It's just that such happenings tend to impress people more.

A psycho-physical union

In any case Jung and Pauli apparently agreed on a final form of their proposed union between psychology and physics. This was a psycho-physical diagram, supposed to show the relation between causality and acausality in everyday affairs.

The Synchronicity diagram then depicts "indestructible energy" – apparently reflected in the physical reality of four-dimensional space-time. From this relation come normal causal connections on one side, and paranormal acausal connections (manifesting as Synchronicity) on the other. Exactly what all this is supposed to mean however, seems very hard to say.

Still some have compared Jung's diagram to a familiar Earth-map with lines of latitude and longitude. So we crawl along those lines of latitude as our life-lines develop, but with our future events (intersections with other life-lines) already predetermined as others crawl along their lines of longitude. And about which future encounters, intuitions operating through Synchronicity, may also sometimes make us aware.

Such interpretations display a certain affinity with the 'Block Universe' – a 19th-century concept which I'll discuss in proper detail in Ch. 18. But whether or not Jung's diagram really meant this now seems difficult to decide. A cynic might even suspect that he wasn't saying anything particularly sensible at all:

> "the idea of synchronicity produces a picture of the world so irrepresentable as to be quite baffling... The term explains nothing, it simply formulates the occurrence of meaningful coincidences, so improbable that we must assume them to be based on some kind of principle"

More importantly in any case those 3 anecdotes I've quoted, can all be interpreted more simply, in our usual terms of pre-call or anti-memory. For these at least there's no further need, for the muddled mysticism of Synchronicity at all….

A Brief Summary

W. Pauli was a renowned quantum physicist, and thought some of its puzzling rules might still be operational, in the sporadic nature of paranormal experience.

He collaborated with equally renowned psychologist C.J. Jung, who had a wide interest in mysticism and psychic matters of all sorts.

Their joint effort in the 1930s produced the notion of Synchronicity, a confusing concoction whose meaning and consequences were neither firm nor clear.

Still their effort was a first attempt to unite the psychological with the physical through the paranormal. And this is a union which anti-memory can now achieve more successfully, as I'll show later in Part Four.

9

REMOTE-VIEWING -CLAIRVOYANCE RENAMED

"This was strong evidence that a viewer looking into the future can describe a picture that he is actually going to be shown at the end of an experiment, rather than the target site itself."
R. Targ, K. Harary -*Mind-Reach* – 1977

A FIFTH TEST for anti-memory is provided by the notion of 'remote viewing', an idea which started in New York in 1971. There, at the headquarters of the American Society for Psychical Research, artist and gifted 'psychic' Ingo Swann was being tested for his intuition capability.

From divining objects hidden nearby, Swann soon progressed to accurate descriptions of people passing outside. And thence to the unlikely statement that it was raining in desert Tucson, 3,000 miles away!

As anti-memory would again predict however, Swann seems to have been mostly, or probably always, informed later about the reality of those true descriptions he had made. For example they rang the weather station in distant Tucson to find out if it was really raining there. (It was!)

So that later confirmation of his earlier intuition certainly occurred in this case. Which is of course just what the anti-memory concept would require.

Though undoubtedly a most gifted intuitor, Swann still didn't seek for reduction of diversity to identity as a scientist might do. (Or indeed as anti-memory can afford!) Instead he believed that his mind had (somehow?) really transcended great distance to view the weather in Tucson, presumably then reporting back to his physical presence in New York!

Clairvoyance renamed

All this was of course very similar to Swedenborg's famous dream of the Stockholm fire (p.49), for which paranormal 'clairvoyance' was generally inferred. Likewise it provides yet another illustration, of that common space-before-time priority we've now met so frequently. But Swann further lived in the modern CCTV (Closed Circuit Television) age. So he coined a new term he thought more fitting to describe his skill. Just how this happened he describes on his website: [1]

> "to be able to put a category of experiments on the pages of reports which were beginning to accumulate, I suggested the term "remote viewing".... since a distant city was, after all, remote from the experimental lab in New York"

'Remote-viewing' was therefore just the old notion of 'clairvoyance' under a modern name. In Victorian times the proven reality of telegraphy had provided a model from physics for supposed 'telepathy'. And now in the electronic era, the reality of television would likewise provide another physics model for the 'remote-viewing' idea!

And so there took root yet another fundamentally mistaken notion of intuition, one destined to cause great confusion and much pointless turmoil over the next 25 years.

At any rate Swann's term was accepted without much question by psi researcher Harold Puthoff. He was an ex-Scientologist and now a laser physicist working at Stanford Research Institute (SRI). As the second biggest think-tank in America, SRI was the research and development arm of Stanford University. Largely funded by military contracts, it employed 3,000 people, on a 13-acre site at the southern end of San Francisco Bay.

Here Puthoff had just been approached with a novel proposal from the CIA. The Agency feared that Russia might be far ahead of America in paranormal mind control, so that a 'mind gap' (similar to the much touted 'missile gap') might result. Wherefore it now wanted Puthoff to look into paranormal capability.

This was the start of a secret government program involving directed application of intuition, a program largely run by physicists for the next 23 years.

From Swann's ready demonstrations, it seemed clear that he'd developed intuition under large control. So his interpretation of it was also adopted quite uncritically from the start. Or as Puthoff's co-worker Russell Targ concluded rather debatably: [2]

> "....remote viewing. The term seemed neutral and free of any prejudgement about how psychic functioning works, and described the data we had seen in the laboratory" [2]

Logicians of course might point out a classic example of loose thinking or *non sequitur* at this point. For even though mediums or intuitors may demonstrate skills to a very impressive degree, it doesn't necessarily follow that they fully understand the same.

Too ready initial acceptance, of this new term 'remote-viewing', therefore left any further consideration of its descriptive verity as quite minimal. (The same thing happened with 'telepathy' back in 1882.) Further the facts would always be forced into a conceptual straitjacket, or theoretical framework, which didn't really fit them at all! (As also happened with 'ESP' after 1930.)

America's official remote-viewing program was then a sort of mini-Manhattan Project for the mind. It was always focussed on paranormal capabilities, which were never really understood. Under various project names and differing bureaucracies, it cost about 20 million dollars spread over more than 20 years.

But in late 1995 – with the Cold War over and all fears of Russian mind dominance well and truly banished – much of this program was finally declassified. So a great deal of information, not all of it reliable, is now available over the Internet. There are also books by some of those involved.

3 of these sources are especially informative right now. First is a series of books by Joe McMoneagle, a former Army Warrant Officer who proved one of the best and longest serving remote-viewers of all. Second is *Tracks in the Wilderness* (Element Books – 1998) by Dale E. Graff, a former director of the entire program. Third is *Psychic Wars* by Elmar Gruber (Blandford – 1999), which ably traces the program's chequered history.

And while these sources don't always agree or inform in finer detail, they still give a good idea of the program overall

In addition there are two earlier books: *Mind-Reach* (R. Targ and H. Puthoff - 1977) and *The Mind Race* (R.Targ and K.Harary - 1984). These books are co-authored by the program's most prominent researchers, and so convey a good idea of its underlying mindset. It's clear that those older space-oriented notions, i.e. *clairvoyance* and *telepathy* were still predominant.

For example the researchers approvingly cite two earlier works - Rene Warcollier's *Mind to Mind* (1948) and Upton Sinclair's *Mental Radio* (1934). These former writers both envisaged 'telepathy' in radio propagation terms. As such both would lend support to this new notion, i.e. that remote-viewing (somehow!) enables one's intuition to surmount great distance and function far away. Like some sort of mental television camera reporting back to base from afar.

From the anti-memory viewpoint, the entire remote-viewing program then grew into a totally misguided, though still partly successful, exercise in directed intuition control. It was misguided because it never properly understood what intuition really is about. But it partly succeeded because people like Swann could demonstrate the faculty under deliberate control.

Not only that, they could easily impart the skill to most other people as well.

Later confirmation – always?

At first these new remote-viewers were mostly people selected because they were regarded as unusually intuitive. For example Ingo Swann was one of them. Another was Pat Price, a former police officer who'd often used intuition in his work. A third was Hella Hammid, a talented local photographer who'd not previously realised how strong her intuition was.

All three worked for years with Puthoff at the Cognitive Sciences Laboratory (CSI) in the Stanford Institute. Initially they operated within the old familiar clairvoyant routines. Much as with the mesmerists two centuries earlier, they employed intuition to divine targets within envelopes, closed boxes, and so on.

Typically the remote-viewer would relax in a quiet room describing or sketching these targets for a second person, a 'monitor' who was always there. The visual or 'viewing' aspect was strongly emphasised, so the sketching was meant to encourage right-brain activity. A neutral 'judge' would then decide how well these sketches, plus any related descriptions, matched up with the reality.

And finally of course the paranormal viewer would be given feedback – i.e. normal information in one form or another about what the target really was. Which again makes anti-memory, the only paranormal process logically required!

Oddly enough also for a program run by physicists, there seems to have been little or no emphasis on direct numeracy. Apparently this was because raw numbers were thought to provide poor targets for the remote viewing skill.

In any case early remote-viewing experiments soon expanded dramatically in range. At first this required another person who would travel to some undeclared spot miles away. Back in the lab the remote-viewer would then use deliberate intuition, trying hard to establish whatever the outsider was seeing at that time.

A good example, of this procedure, is provided by Joe McMon-eagle. For his first remote-viewing session he spent 18 minutes concentrating strongly, trying to gain and integrate intuitions about the unknown spot where his two outward-bounders were. Slowly his thoughts cohered into an organised impression of a building, which he then sketched out.

About one hour later he was taken to see the Stanford Art Museum, the undeclared scene where the outward-bounders had been and which he'd been trying to visualise: [3]

> "It is very difficult to explain the type of excitement one feels when you suddenly see your imagination displayed before your eyes.
>
> "I had never been to the Museum before. But at that moment in time, I knew that it was the very building I had laboriously constructed in detail within my mind.
>
> "I found that I wanted to kick myself for having not paid attention to all of the minor detail"

Later Swann realised that having some other person at distant sites was quite superfluous. This was a conclusion which Puthoff and Targ found difficult to accept at first. Still soon it proved sufficient to describe the target location by its geographical co-ordinates alone. Later it emerged that these too could be jettisoned, so that any form of random labelling would do.

All of which would of course have followed quite automatically from the anti-memory concept!

But by now the notion of 'remote-viewing' had aroused CIA interest in its spying possibilities. The Agency then funded Project Scanate between 1973-75. Its object was to spy into Cold War 'hot spots', identified mainly by geographical co-ordinates.

Under Scanate Pat Price drew a good visual impression of a secret Russian atom site at Semipalatinsk, an impression which satellite photographs soon confirmed for him. But he also described thick steel segments within the building, a fact not confirmed until well after he died suddenly in 1975.

Unlike almost all other remote-viewing episodes, this last feat would certainly not make sense in the usual anti-memory terms. So perhaps it involved some higher order of mystery, much as my own personal analyses seem to have encountered sometimes.

In any case at a later stage Dale Graff became director of the remote-viewing program, now renamed 'STARGATE'. As a man with strong personal experience of intuitions, many related in his book, he was unusually well fitted to realise that the remote-viewers were on to something real. Still again he exhibits that usual 'space-before-time' priority of thought: [4]

> "Psi scanning required relinquishing a hold on only one of the features of the universe – space. (But to relinquish) the other feature of the universe - time – seemed to be much more improbable"

In response to which one might argue that the physical laws of space have been probed by science for over 2,000 years. And they seem to offer no niche whatsoever, for anything like remote-viewing in any form. In contrast to which the laws of time – assuming of course that there can be any such! - are as yet quite mysterious and unknown. So that we can reasonably expect strange, unexpected, findings from them in due course!

In any event the predominantly spatial notion of remote-viewing, suggested various experiments to investigate effects like shielding and so on. Graff cites one such where his group was deep inside the Ohio Caverns – and so well shielded from low-frequency radio waves! Still they were successfully remote-viewed by Hella Hammid, who was 600 miles away.

However of course and as usual, she was presented with the details shortly afterwards! Once again therefore anti-memory would have been quite sufficient, to interpret this episode!

Later too Swann and Hammid were further successful from a submarine under San Francisco Bay. But, again as usual, they were debriefed on resurfacing about the true nature of their targets. Still, *"Salt water provided no attenuation effect"* as Puthoff noted soberly...

Military tasks

With thinking and findings of this order, the U.S. military soon became involved. This led to a special unit named Project Grill Flame being set up in 1978 at Fort Meade in Maryland, under Major Gen. Ed. Thompson. He wanted real-world applications immediately, leaving it to the scientists to explain eventually: [5]

> "We didn't know how to explain it, but we weren't so much interested in explaining it as in determining whether there was any practical use for it."

To this end new remote viewers were now recruited, mostly through personal acquaintance combined with a psychological profile. And even though just about anyone could apparently succeed at this particular variant of controlled intuition, photo-interpreters were naturally expected to excel.

A total of perhaps 18 army personnel was employed operationally in this way. There was a similar number concerned with more scientific pursuits at the Cognitive Sciences Laboratory. Joe McMoneagle therefore estimates there were never more than three dozen remote-viewers in all. There were also some 50 other staff employed at various times, about equally divided between scientists and support personnel.

At Fort Meade the Army viewers relaxed in small grey rooms in wooden huts, trying to will their intuition "through the ether" into Cold War locations far and wide. For example in 1979, McMoneagle remote-viewed that a huge submarine was being constructed, inside a large shed by the sea in Northern Russia.

This was duly confirmed for him 4 months later, when spy satellites spotted Russia's first Typhoon submarine being launched from there.

Other remote-viewing missions however proved more problematical. For example around early 1980 hundreds of sessions failed to locate the Iranian Embassy hostages. But again from the anti-memory viewpoint, this failure must seem quite expected and predictable. For there were never any satellite photographs, or other forms of later confirmation, to be presented afterwards.

In any case one crucial aspect of remote-viewing was by now well established – the fact that no unusual 'psychic' abilities were required. Apparently just about anyone could learn the skill. Though some would always be better than others, as indeed with just about every other feat of mind.

Successful control of intuition in this manner could also be greatly degraded by untoward circumstances neatly termed 'inclemencies'. Broadly these were the sort of thing which can impair any other sort of high intellectual function – factors like hangovers, allergies, colds, mental stress, annoyance, etc.

All of which further confirms my own impressions about anti-memory or pre-call skill. Apart from its novel time orientation, it's much like any other high-concentration exercise.

End of a program

By the 1990s however the Cold War was obviously winding down. So official support for remote-viewing began to wax and wane, dependent on bureaucratic office politics. Some viewers had become destabilised from their efforts to "see through the ether", and most of the original researchers were gone. [6]

In July 1995, the CIA was therefore requested by Congress to resume control of their former STARGATE program. But the Cold War was now well and truly won in the western interest, and the Russians had never made their once feared breakthrough into mind control. A new regime at the Pentagon was also more sceptical of the entire remote-viewing program, doubting if it had ever produced any worthwhile intelligence at all.

Before it would consent to take charge of the program again therefore, the CIA commissioned an independent study which became known as the AIR Report.[7] It's unclear why AIR apparently only considered the last two declining years, some 10% of the entire program. This hardly inspires much confidence in its conclusions, as nor does the fact that they were reached after just 3 months, in September 1995:

> "Even though a statistically significant effect has been observed in the laboratory, it remains unclear whether the existence of a paranormal phenomenon, remote viewing, has been demonstrated.
>
> "The laboratory studies do not provide evidence regarding the origins or nature of the phenomenon, assuming it exists…the information provided by remote viewing is vague and ambiguous"

This conclusion was echoed by Oregon psychologist Ray Hyman who found that remote-viewing was about 15% accurate – which he ascribed to informed guesswork. In this he was challenged by former Director Ed May who claimed 50% accuracy. May is in agreement with star remote-viewer Joe Mc Moneagle who estimates around 55% accuracy in his latest book. Both are supported by statistician Jessica Utts, who served on the same review team as Hyman:

> "Using the standards applied to any other area of science, it is concluded that psychic functioning has been well established.
>
> "The statistical results of the studies examined are far beyond what is expected by chance….
>
> "(They present) several examples of prima facie evidence, i.e. evidence having such a degree of probability that it must prevail unless the contrary be proved"

Most sensibly it then seems that genuine intuition effects were indeed being generated continuously within the various remote-viewing programs. But they never provided much firm intelligence that could be utilised. So in practical terms the CIA wanted rid of it - and provided a rather incompetent piece of window-dressing as its excuse.

Still the Agency did let it be known that it might see fit to work from time to time with private contractors on similar concerns. As indeed it's probably doing still today.

Anti-memory explains

What then can one make of remote viewing overall? That it was the most extensive ever investigation into the nature of intuition – both theoretical and practical – is quite clear. But that it also always involved three basic errors is also unavoidable.

The first mistake was failure to note *all* of the basic facts, about Swann's successful intuitions in those original New York experiments. For example when he divined the weather in distant Tucson, he ascribed little or no significance to his near-immediate telephone confirmation of the same.

Secondly, from this incomplete appreciation of *all* relevant facts, came Swann's notion that he was somehow 'remote-viewing' those distant sites. Instead of just merely pre-calling relevant information, that he would receive at a later time.

Thirdly the Stanford researchers accepted such loose thinking and terminology, without further consideration and too readily. (They should have employed a good historian or philosopher of science, for more critical evaluation of their theme!) So that this third failing condemned many volunteers, into some hopeless tasks and much needless confusion, over too many years.

The program however did mean a large step forward in the deliberate control of intuition, establishing also that just about anyone is capable of the feat.

Otherwise nearly all the evidence now available suggests that the remote viewers were mostly, more likely always, just unwittingly using anti-memory throughout. This is because feedback was apparently provided after most, perhaps all, successful trials. Or in other words the paranormal intuition was usually followed, by later normal confirmation in the usual way.

And whether or not many trials without feedback were ever successful is a question we must leave to others, those with more access to the records to decide.

It's also more simple to read anti-memory into most or all those individual incidents which remote-viewers have reported. For example Dale Graff's book is a valuable repository of many such intuition anecdotes. But inevitably every last single one of them makes better sense, in terms of anti-memory at work.

The same is true of Joe McMoneagle's refreshing and candid accounts in his several books: [8]

> (McMoneagle remote-viewed) a red bicycle, in a cycle rack near the front door of the target building.
> However the 'out-bounder' person didn't report this - because there was no bicycle there at first.
> But when they went back to inspect the target later for feedback, somebody rode in on a red bicycle, then parking it as McMoneagle had foreseen.

Now this little incident of course makes perfect sense in simple anti-memory terms. It also fits well into the pre-call diagram. For that red bicycle event was one of unusual interest - and so likely to project strongly in either memory mode.

All of McMoneagle's books likewise convey obviously genuine accounts about 20 years' practice with intuition under deliberate control. But if one were to substitute the term 'pre-call' every time he uses the composite verb 'remote-view', each separate account becomes much more comprehensible overall.

Finally to illustrate my main point here, there's a very simple 'remote-viewing' experiment in which any reader can readily participate right now.

Just close your eyes and try to re-call the details of some far-off place, which you may have visited last week or year. These details will come back to you in bits and pieces, as you gradually *re-construct* a mental vision of that distant scene. Indeed you could even sketch your impressions of it – though these will hardly prove fully accurate, if you then compare them with some actual photograph!

But it would surely be silly for anyone to consider your feat as 'remote viewing' of that now far-off place. It's simpler and more sensible to ascribe your exercise to memory alone.

This little mental exercise which you've just conducted, is also of course all very similar, to what the remote-viewers used to do. The only real difference is that you employed normal time orientation towards your own *past* experience - while they had a paranormal time orientation towards personal *future* experience.

Or in other words you were trying to *re-call* while they were trying to *pre-call* – two very similar mental exercises just diametrically opposed in time.

Wherefore there seems no logical need to consider controlled intuition in terms of some mysterious remote-viewing faculty. It's simpler to think that the Cold War warriors were just using anti-memory throughout. So that while they imagined they were transgressing *space* paranormally, more likely they were just transgressing *time*. Their victory lay in the fact that they were rather good at what they were doing, their failure in that they never understood their own actions properly.

But in any case anti-memory or pre-call can again explain just about all of those feats ascribed to 'remote-viewing' – and so survives this fifth test as before.

A brief summary

Remote-viewing was the old notion of clairvoyance, updated by inference from CCTV technology into a more modern term. Between 1971-95 this term then led some members of U.S. intelligence circles, to project that intuition could be used, for spying into enemy places far away.

The resultant program was by far the most intensive effort at intuition control ever carried out anywhere. And it showed that many or most people could achieve this skill. But it failed to provide much firm military intelligence, and so was declassified in 1995.

All public accounts made available since then, are amenable to the usual interpretation of pre-call or anti-memory. This possibility never seems to have been realised throughout the entire program. But still it appears to have been the one single process logically required throughout.

As such this whole episode provides yet another illuminating example, of that common space-before-time priority in most thought.

10

AN EXPERIMENT WITH TIME

> *"Here is a piece of knowledge concerning which the blind man had no previous conception!"*
> J.W. Dunne - *An Experiment with Time* – 1927

MY RESEARCH INTO INTUITIONS has therefore revealed one very consistent pattern of thinking, over the past two centuries. In five main episodes - *'clairvoyance', 'telepathy' 'ESP', 'Synchronicity', 're-mote-viewing'* – the investigators postulated some mysterious mental power supposed to function without limit across distance or *space*. While largely or wholly ignoring any similar possibilities for *time*.

Dunne's prophetic dreams

A sixth and final major test for anti-memory is however provided by one paranormal investigator who failed to follow the common trend, and so made a major contribution to the field. He was John W. Dunne (1875-1949) born to a military family in Roscommon, Ireland. Curiously this lies just 40 miles north-east from Galway where I conducted my first Intuition Surveys.

Dunne grew up as a military man and inventor, involved in developing British warplanes before the First World War. And in 1927 he published *An Experiment With Time,* a justly famous book which has mostly remained in print ever since.

In this book Dunne relates a total of 35 intuitions he had experienced over many years. Mostly these were intuitive paranormal dreams - for which he was inclined to favour telepathy or clairvoyance at first. But then his thinking was clarified by an especially significant dream intuition, one where a curious correlation between numbers clarified what was really going on.

The year was 1902, when Dunne was with the British Army in South Africa. There he dreamt he was on an island about to blow up, like Krakatao some 20 years before:

> "Forthwith I was seized with a frantic desire to save the four thousand (I knew the number) unsuspecting inhabitants.

> "There followed a most distressing nightmare, in which I was trying to get the incredulous French authorities to remove the 4,000 inhabitants of the threatened island.
>
> "All through the dream the number of the people in danger obsessed my mind. I repeated it to everyone I met".

In due course *The Daily Telegraph* was delivered out from Britain, confirming the content of this scarifying dream. It told how Mont Pelee on the Caribbean island of Martinique had exploded without warning, completely devastating St.Pierre, the capital. The *Telegraph* headline proclaimed the "probable loss of over 40,000 lives" – a figure which later turned out to be a large over-estimate:

> "The true figure had nothing in common with the arrangement of fours and noughts I had dreamed of.
>
> "So my wonderful 'clairvoyant' vision had been wrong in its most insistent particular! But it was clear that its wrongness was likely to prove just as important as its rightness.
>
> "For whence, in the dream, had I got that idea of 4,000? Clearly it must have come into my mind because of the newspaper paragraph."

Or in other words Dunne hadn't really dreamt of the actual event – but rather the headlines he would read a few days afterwards! For him this cut out all need for space-violating notions like 'clairvoyance' or 'telepathy'. Henceforth therefore he would focus totally on time transgression alone.

There was still the possibility of déjà vu or identifying paramnesia, two forms of mental confusion about sequence in time. So Dunne took to writing down the details of any dreams he could remember immediately on awakening. This had to be as soon as possible, or else he would forget them rapidly! Then he kept watch for any correspondence over the next few days: [2]

> Dunne dreamt he was on a planked walkway visible through smoke, crowded with people who dropped in heaps. They were choking and gasping as snakelike objects threatened them through the gloom.
>
> The complement came with the evening papers, which told of a fire in a Paris rubber factory. The women workers had crowded out on a balcony, where they all suffocated from the choking fumes.The fire ladders and hoses were too short to reach up to them!

Not all of Dunne's dreams reflected events like these, recounted in newspapers he would later read. What impressed his readers most was that so many of his dreams concerned the everyday or commonplace.

For example he dreamt of a secret loft, which he used to escape from a house – and next day encountered a novel with this theme.

Other Dunne dreams involved some personal experience he would encounter later, and mostly with that little extra touch of the unusual or bizarre:

> Dunne dreamt of a huge horse pursuing him in a threatening way.....Out fishing with his brother next day, he found himself watching a horse, much smaller, in a field across the river, rearing and plunging as in his dream.
>
> He was just relating the dream to his brother when the animal somehow escaped. It went thundering down the path to those wooden steps Dunne had seen the night before, swerved past them, and then plunged into the river towards its audience of two.
>
> They picked up stones to defend themselves. But the horse merely snorted derisively at them, before galloping off down the road!

An Experiment with Time recounts 35 such incidents in all. Each and every last one of them fits easily on to the pre-call diagram (Ch. 4). So they can all be interpreted in terms of a weak anti-memory faculty not always accurate in full detail. For example that crucial confusion - between 4,000 and 40,000 in his volcano sequence - is just the sort of mix-up you might expect with normal, past-oriented, memory.

Serial Time

However Dunne never saw things in this simplest way. Earlier he had rejected telepathy and clairvoyance (both of which he'd once believed in!) in favour of foreknowledge alone. But now he strayed off from that most productive path of maximum simplicity, describing the basic facts of his intuitions in a manner that was needlessly complex.
He thought there were 4 essential elements to his dreams:

a' – a *pre-presentation* of A (e.g. his dream of those newspaper headlines)
a" – *a re-presentation* of a' (i.e. his later memory of the dream)
A – a *presentation* (his actual reading of those later headlines)
a – a *re-presentation* of A (his later memory of the same.)

Now all of this seems to me to be two complexities too far. To describe the full facts with maximum economy, you only need a' and A. That is to say a dream of those headlines, and a later reading of the same. In which case too, the concept of anti-memory again suggests itself almost automatically.

At this crucial point therefore Dunne diverged from simplicity to produce a rather baffling theory, which he termed Serial Time. He proposed that your mind has an infinite layer of different observers - Self-1, Self-2, Self-3, etc. Self-2 can observe both past and future of Self-1 as the latter moves through its own time in everyday experience. Self-3 can observe Self-2 likewise, and so on.

Sometimes too the latest Self permits the previous Self brief glimpses of the latter's future - through intuitions and prophetic dreams.

But with this complex theory Dunne lost a lot of his followers who supported him on intuition: they couldn't really understand what he was now on about. In addition philosophers like C. D. Broad rejected Serial Time because that endless procession of different Selves meant an "infinite regress" – a procedure basically forbidden in philosophy.

More importantly in any case, Dunne was never really able to bring his dream intuitions under more conscious control. Nor indeed offer any further proof, beyond his advice to keep strict watch on your own dreams each morning as you awake.

Still Dunne found a fan in J.B.Priestley, the famed English author with a strong interest in the temporal. In his book *Man and Time* (1964) the latter observed that most scientists really know very little about time. Then he proceeds to consider Dunne's contribution in large detail.

But otherwise Dunne was largely ignored by the paranormal researchers of his era, notably J.B.Rhine and the SPR. Likely this was because he'd discarded their main tenets of space transgression - 'clairvoyance', 'telepathy', 'ESP' – as logically redundant or unnecessary..

Nevertheless today Dunne occupies a unique niche in the record of intuitions overall. He's probably the only investigator whom conventional science treats with some circumspection or indeed respect. He's also the only one ever mentioned in serious time studies elsewhere. For example, in his definitive *Natural Philosophy of Time,* the authoritative time scholar G.J. Whitrow gives Dunne two pages, suggesting that perhaps after all Dunne really did have something worthwhile to say.[1]

In a real sense also this book you're now reading follows in Dunne's footsteps to a degree. For of all those earlier investigators into paranormal intuition, he's the only one who ever resolved the basic problem into terms of just time alone.

de Pablos and Dunne dreams

In recent years too Spanish researcher F. de Pablos has followed in Dunne's footsteps by monitoring his own prophetic dreams.[2] Each day for 5 years he recorded all those dreams he could recollect, thereafter searching systematically for any evidence of time violation in them.

He found such evidence in 124 apparent dream intuitions over the years 1996-2000. Then he used his own personal experience of their manifestation, to decide between two different models of what may be going on.

The first, often postulated, model is Signal Transmission Theory (STT). It postulates transmission of information *backwards in time* from the future to your present Mind. This might be through tachyons, which are still theoretical swifter-than-light particles. Or it could happen through 'advanced' light waves, as I'll explain in Ch. 18.

If this were so however, one must expect the sequence of dream intuitions to occur, in the exact same order as their real life counterparts. For example if a wedding on Wednesday is followed by an accident on Thursday, then by STT an intuitor dreaming beforehand must always 'experience' some semblance of the wedding before the accident. For him to do otherwise would require light signals carrying signals 'back from the future' to travel at different speeds, a variation which physics entirely forbids.

In practice however De Pablos finds that future-oriented dreams simply don't concur with real-life sequence. So people may dream about the accident *before* they dream about the wedding, just as often as in the reverse sequence.

De Pablos therefore holds that STT is impossible, and finds that the evidence favours a model based on Episodic Memory instead. Episodic Memory is a particular sector of the more general memory process. It's the one that deals with our conscious recollection, our re-call of those past events we once experienced personally.

With Episodic Memory of course there need be no strict congruence with the actual sequence of real-life events. So you can re-call what you've just had for dinner at one moment, and some incident from 20 years ago immediately afterwards. Or indeed reverse this sequence at your pleasure - only considering your dinner after your contemplation of so long ago.

Dream intuitions likewise don't occur in any strict sequence, or in congruence with the actual order of events. De Pablos therefore concludes that, in preference to STT models, they're really most similar to conventional - i.e. past-oriented - personal memories.

In addition, as I've earlier reported (Ch.4), de Pablos finds a strong correlation pattern in time-lag – the interval between a dream intuition and its apparent outcome in real life. Of his 124 dream intuitions therefore, 60% were 'fulfilled' within the first 24 hours, 14% over the next day, and so on. 19 intuitions (15%) appeared to concern events between 4 and 25 days ahead, and the longest interval was a full year into future time.

All this of course substantiates my own analysis of intuitions as I've described in Part One.

There are however four main differences. Firstly de Pablos analysed dream intuitions only, while I was mostly concerned with the wide-awake kind. Secondly he doesn't extend his analogy to cover all other putative manifestations of the paranormal – *'clairvoyance', 'telepathy', 'ESP',* etc. Thirdly he hasn't reported intuition under more direct or wide-awake control, as I'll be considering in Part Three.

That said however de Pablos has analysed his personal intuitions - and substantiated their analogy with personal, past-oriented, memory - in very large detail. Overall his approach largely complements my own, one being specific and the other more generalised. Together they strengthen the pre-call or anti-memory hypothesis to a new degree.

Dreaming of races not yet won

Other people too continue to have Dunne dreams, sometimes gaining large media attention through their unusual content. One such was John Godley, the future Lord Kilbracken from Cavan in Ireland. At the time he was newly demobbed from the British Fleet Air Arm after World War Two.

Godley then became a mature student at Balliol College in Oxford University, where he started having occasional dreams of horse races which hadn't yet taken place. These he described in his book *Tell Me the Next One* (London - 1950),

I'll deal with this minor episode, of gambling based on intuition, at some length. This is because most of its carefully reported secondary details reflect my own experience, and further show parallels with conventional past-memory to a new degree. Here the crucial point is that Godley seldom dreamt either the actual names of the horses, nor their winning prices, with full accuracy: his dreams never quite corresponded to the reality. As of course you would expect from any weak memory capability, in either past- or future-oriented mode.

Over 3 years Godley then reported 10 winning horse dreams. Of these he bet on 9 races – and actually won 8 outright. Another horse was placed 3rd though he had no bet on it, and there were special circumstances for the final one! These dreams therefore resemble Dunne's similar horse sequence that also came true, the great difference being that Godley's were always focused on races where money could be won.

Whenever Godley dreamt of a horse race he would tell some of his fellow students, and then check up in the newspaper racing columns. If some horse corresponding to his dream were thus detected, he would back it with fairly respectable sums, equivalent to about 50 pounds or 100 dollars today.

So that, even though their focus was rather different, Godley's dreams further resembled Dunne's apprehensions of newspapers as yet unread.

On the night of March 8, 1946, Godley first dreamt very vividly that *Bindal* and *Juladin* – two horses he'd never heard of - had both won at 7/1. Along with some friends he checked the papers, to find *Bindal* running at Plumpton, and *Juladin* at Wetherby.

The former duly came home, at much reduced odds of 5/4, the latter at a more respectable 5/2. His dream was therefore doubly correct though not fully accurate, again supporting my more general finding that anti-memory is seldom correct in all details.

For Godley's next dream he was on vacation, in the Kilbracken family seat in County Cavan, Ireland. Once again this is curiously just 50 miles northeast from Roscommon, where Dunne was born in the previous century.

There, on the night of April 4, Godley dreamt that *Tubemore* had won the following day. A horse called *Tuberose* entered for The Grand National was the nearest name equivalent. So a small local circle who were in on the secret had it well backed when it duly romped home at odds of 100/6 or 16.7 to 1.

On July 28 the same year Godley dreamt that *Monumentor* had won at 5/4. A check with the papers revealed that *Mentores* was running; as before it duly won for him at odds of 6/4.

One year later – on June 13, 1947 - Godley dreamt of a winning double, though now in two very different ways. In his dream he was at the track and heard all the punters shouting the name of the favourite which was *The Bogie*. He discerned no name for the other horse - but recognised the face of the jockey Edgar Britt, in the distinctive colours of the Gaekwar of Baroda (an Indian racing potentate).

When the morning papers showed *Baroda Squadron* running at Lingfield and *The Brogue* as a firm favourite, Godley knew exactly what to do. After telling various people before the race, he wrote a statement, had it time stamped at the Post Office, and then had it locked up in the Postmaster's safe.

Then he rang *The Daily Mirror,* a newspaper that proved very interested. And when both horses won (at the rather miserable odds of 11/10 and 1/9 respectively) fame came flooding in.

Godley now took up *The Daily Mirror's* offer to be a racing correspondent – and soon produced his first failure! On Oct 29, 1947 he dreamt of *Claro* – which he backed but was unplaced. But on Jan 16 1949 he dreamt of *Timocrat* which won at 4/1. Three weeks later he predicted, but didn't bet on, *Monk's Mistake* which came third at 6/1. On the same day he bet on *Pretence* which won at 8/1.

But while working for the newspaper he felt that his strange gift was much weakened, as indeed usually happens with intuitions when any element of stress intrudes.

John Godley however had one final winning episode. 9 years later he was the new Lord Kilbracken in the family's Cavan seat. Just as in 1946 he dreamt of another grand National winner on the night before the race. His new dream horse was now called *What Man* – for which *Mr. What* was the nearest counterpart in reality. He was backing it for a large sum (unspecified) when it duly romped home first.

Godley's story of dreaming race winners resembles many others similar scattered throughout the intuition records. It differs mainly in that he reports a whole series of the same. It's important in that his dreams of those future winning horses were seldom accurate by name.

This is of course what one might expect from a weak memory faculty in either past or future mode. It would seldom be totally correct in all detail!

Premonitions of disaster

The same is true of those many cases of seeming foreknowledge, apparently focused on the Welsh Aberfan disaster of October 21, 1966. There a huge water-weakened coal tip slid down in an avalanche of black slurry on the little mining village, killing 144 of its inhabitants. Of these 128 were children at the local junior school.

Dr. John Barker was a London psychiatrist who traveled to help at the disaster scene next day. Already much interested in matters psychic, he realised that this incident might provide a classic source of intuition anecdotes. So that a national media appeal for premonitions was circulated within the week.[3]

In all about 200 people – or 4 per million in Britain's total population – claimed to have had some inkling of Aberfan before the tragedy occurred. Barker investigated 76 of these claims in detail, ascribing significance to 35 of them. Of these he found 24 to be 'confirmed' – i.e. related to someone else before the tragedy occurred.

One of the most touching anecdotes concerned little 10-year-old Eryl Mai Jones. She was to die in the disaster and mentioned it to her mother twice before:

> "Mummy, I'm not afraid to die.....I dreamt I went to school and there was no school there. Something black had come down all over it!"
> And yet she skipped off to school happy as usual on the disaster day!

Barker reported that the time lag, or interval between premonition and disaster, ranged from one year down to a couple of hours. The premonitions, over half of them as Dunne dreams or nightmares, tended to multiply as the event drew near.

The original intuition or premonition was of course always confirmed later, mostly by the TV news. But in no case – excepting little Miss Jones – were these intuitions fully specific or accurate.

Barker's investigation of Aberfan intuitions remains unique because they're all so unequivocally oriented into future time. His findings were broadly like Dunne's earlier conclusions, though with a much wider population spread. And as for those other notions of clairvoyance and telepathy, they simply could never have applied here.

Rather, as with all other accounts in this chapter, these Aberfan anecdotes all very definitely indicate foreknowledge alone. This is that clear common factor I had originally analysed behind intuitions of all sorts (Part One). And, as my analysis also requires, later confirmation very obviously happened in every last single incident.

A supporting secondary feature is that the Aberfan intuitions seem to have grown more frequent as the actual event drew near. Again this supports the conclusions of de Pablos I've reported above, and also my usual comparison with ordinary memory. For in normal experience also, our memories naturally maximise just after some incident has occurred.

But neither are these ordinary past-oriented memories ever totally correct in all detail.

So that throughout this final sixth test also, the concept of anti-memory again holds true. It's the simplest proposal yet advanced, to cover the endless variety of intuition anecdotes.

A Brief Summary

Historically speaking, J. W. Dunne in the early 20th century was the only major figure ever, to have considered intuitions exclusively in terms of time alone. He focused on over 30 of his own outstanding dreams, which seemed to him clearly concerned with events to come.

From these he proposed a new complex Theory of Serial Time. But he never achieved more direct intuition control outside the dream state.

Other people have also reported Dunne dreams, notably F.de Pablos who has greatly developed the parallels with personal memory. Similar dreams have been reported for horse races and the tragedy of Aberfan. In content and detail all these can again be explained in terms of anti-memory, so strengthening the original hypothesis still more.

11

CONCLUSIONS CONFIRMED

"Theories typically arise as a result of multitudes of observations, in the course of which a deliberate attempt is made to sort out the wheat from the chaff"
Murray Gell-Mann – *The Quark and the Jaguar* – 1994 – p.77

SO ENDED MY LONG TRAWL through the literature of the paranormal, an effort that felt like wading through endless acres of verbal sludge at times. For when people uncritically accept wrong words as their basis for thinking, and so become misdirected into fruitless efforts and sheer nonsense, it can be painful to follow them into conceptual cul-de-sacs, which reason says must lead nowhere.

Nevertheless my long trawl, through so much poorly reasoned literature, had largely confirmed those conclusions I'd first reached personally:

1/ Not all intuitions can be interpreted in purely spatial terms. At least one-fifth involve clear time transgression alone.
2/ But virtually all intuitions can be interpreted in time transgression terms.
4/ *If* space transgression were still really valid on rare occasions, it was best set aside for the moment, as perhaps indicating some higher order of mystery.
5/ Meantime time transgression provides a more reasonable option nearly always.
6/ Considered therefore in terms of time only, intuitions are best *described* as incidents of pre-call or anti-memory.

To which of course most people's first or instinctive reaction might be that anti-memory is obviously impossible. "Surely the nature of time forbids it!" would be the familiar retort, or knee-jerk response.

But in reality the true situation is by no means that simple. For one thing the "true nature of time" is one of the most intractable problems known to science and philosophy. For another, as we'll see in Part Four, anti-memory might almost appear required by the seeming real nature of time, as suggested by Einstein Theory of Relativity.

At this point however I'll just pause to summarise the last 6 chapters. They've dealt with the 6 main episodes of intuition research over the past 3 centuries. With each I will summarise how the anti-memory idea was always a simpler alternative, and further clarify any other lessons relevant....

Space before time.
Throughout most of these 6 main episodes of failed research, there was always a strong tendency to think in terms of space before considering time. This striking 'space-before-time' priority is of course one well known to time scholars, like J. L. Synge or G. J. Whitrow, and may emanate from that first sequence in which babies learn to comprehend reality.

It's also agreed that science has largely developed with the same priority. And the same holds true for nearly all speculations about the paranormal, whose start we can date conveniently to Swedenborg's mysticism 250 years ago.

Gravity and magnetism – both of which clearly transcend space – were much to the forefront at that time. Mesmerism or 'animal magnetism' - which we now call hypnotism - apparently worked likewise. So that when some 'mesmerised' subjects showed heightened intuition, it was natural to infer similar space-transcending mental powers. Hence came the French notion of *clairvoyance* – the supposed ability to see or know about objects out of sight.

Yet most or all records of *clairvoyance* can alternatively be interpreted, in pre-call or anti-memory terms. This is because the paranormal intuitor, would naturally have attained normal confirmation, at some future point.

After the French notion of *clairvoyance* had failed to develop over a century, the British Society for Psychical Research (SPR) put forward the notion of *telepathy* around 1882. Supposedly this involved 'sender-transmission-receiver' as its psychological components. Such notions are similar to the physics of telegraphy introduced not long before. Further they seemed reinforced when radio was discovered a few years afterwards.

Nevertheless the SPR records of everyday intuitions are numerous, well researched and detailed. And so 109 of the best cases were soon advanced as evidence for telepathy. However all these anecdotes - and with no exceptions whatsoever - display later normal confirmation by the intuitor, of some earlier paranormal thought. As such they again conform to the anti-memory profile, and further the more specific pre-call diagram.

With *telepathy* having failed to progress over the next 50 years, the wider but more confused notion of *Extra-Sensory Perception (ESP)* was promulgated in America by J.B. Rhine. He conducted numerous card-guessing experiments, which provided strong evidence for paranormal function at first. Later however, this striking initial success died totally away.

This striking decline in paranormal function may have stemmed from a crucial change in personality, magnified by a further change in Rhine's card-guessing routines. In the early, and most successful years, these seem to have involved *immediate* later confirmation after each guess was made.

But then this routine was changed to a different Down-Through procedure, where later confirmation was withheld or delayed. And as anti-memory would rationally predict, card-guessing success then largely died away. (Though more detailed research is required on this point.)

Conversely later confirmation is standard in *Ganzfeld* procedures introduced 30 years ago, and this is the most successful form of parapsychology experiment to date.

Rhine also received numerous accounts of everyday intuitions, and 186 of the best were published eventually. Though these were all construed in terms of ESP, 179 of them can equally be interpreted as instances of anti-memory.

Meantime in the 1930s, Swiss psychologist Carl G. Jung dabbled in mysticism, with various episodes of intuition prominent. On these he collaborated with physicist Wolfgang Pauli, to produce the confusion of *synchronicity*.

Once again however all of Jung's intuitions can be interpreted more economically, in the usual anti-memory terms. Still his collaboration with Pauli was a notable first attempt, to bridge the great gap between physics and psychology, through new understanding of intuitions in everyday experience.

Remote-viewing was yet another interpretation of intuition advanced in America, and government-sponsored for over 20 years from 1971. Though largely misdirected through faulty observation and poor understanding, this was always the most intensive and nearly successful effort at intuition control so far.

But if you just substitute 'anti-memory' every time 'remote-viewing' is mentioned in its various publications, a massive simplification – or a great gain in clarification - is attained. The same applies to those many anecdotes now available, nearly all being easier to comprehend in pre-call or anti-memory terms.

Indeed just about the only major investigator ever, to have avoided the usual space-before-time priority, was English inventor J.W. Dunne during the early 20th century. Significantly too he was almost the only major research figure ever, to have claimed much personal intuition experience. Dunne's book *An Experiment with Time* (1927) details 35 intuitions, mostly from his dreams. These he found complemented by reality – though never in full detail – at some future time. He tried to explain them through his complexity of Serial Time, not realising that anti-memory could have dealt with them all more effectively.

More recently (2004), Spanish investigator F. de Pablos has investigated his own dream intuitions, and in much finer detail than Dunne ever did. In several aspects he finds they resemble a time-inverted version of personal or Episodic memory. This concurs with the wider anti-memory concept once again.

Totals summarised.

In sum therefore all those 6 major historical interpretations of intuition can now be encompassed, within the single unifying framework of anti-memory. There's simply no logical need to invoke rival space-violating notions like *'telepathy'*, *'remote-viewing'*, *'ESP'*, etc.

Finally too all those matters we've just considered, can now be quantified to a first degree. Though my totals here only sum up, those samples I've listed in the various chapters concerned.

Nevertheless this summary of the literature also confirms my original conclusions derived from purely personal experience. So that virtually all (331/338 = 98%) published intuitions are best regarded as cases of time violation alone.

Chapter	Traditional Idea		(or) ANTI-MEMORY
5	Clairvoyance	4	4
6	Telepathy	109	109
7	ESP	186	179
8	Synchronicity	4	4
10	Serial Time	35	35
Total		338	331

TABLE 11.1: Nearly all intuition anecdotes in the literature of the paranormal, can be unified in terms of just pre-call or anti-memory.

This finding then provides objective proof that my own first 2 surveys of intuition were both representative and reliable. It confirms my original conclusions derived from purely personal experience.

In fact the percentage agreement is again so close (ca. 98% in both cases) as to be remarkable of itself. The simplest conclusion, from this agreement, is that we've really been dealing with genuine intuitions – and with minimum chance input - throughout. But in any case their total affords a uniquely firm basis from which wider conclusions may be generalised.

A repressed faculty

There is in any case more to be gained from the paranormal literature. It contains many other secondary observations supporting my own personal conclusions still more. These secondary features surface as impressions, expressed by proven intuitors like Mary Craig, Gilbert Murray, and the more recent remote-viewing community. All of them agree that intuition is always a labile, delicate, shy and elusive faculty.

In this once again they confirm my own conclusions based on personal experience. It therefore makes sense to combine all these impressions from whatever quarter. This gives a clearer picture of intuition as viewed from the most coherent anti-memory viewpoint.

For this purpose we can ignore the *present*: it's already long *past* as far as physics is concerned. To reconsider a homely example I've already given, suppose somebody trod on your big toe while you were looking elsewhere. Since the speed of nervous transmission is only about 100 metres per second, or 200 miles an hour, it could then be .02 seconds before you realised what had occurred. *"What we mean by 'right now' is a most curious thing!"* as physicist Richard Feynman once said.

The crucial question is then whether Mind or memory must really be confined to the past alone. Or do we all have a similar if rudimentary faculty for anti-memory? - one basically oriented towards the *future* as intuitions indicate. A faculty perhaps left largely latent, or even highly censored, ever since infancy?

Several slight straws in the scientific wind support this latent possibility. One is the reply from pioneering Swiss child psychologist Jean Piaget to a query from Einstein. The physicist had asked the psychologist whether our everyday notions of time are really innate and obvious? – like the difference between heat and cold. Or might they instead be a *learned* achievement? - like the ability to ride a bicycle or speak Chinese.

Piaget then studied the development of time awareness among the very young to conclude quite unambiguously: [1]

> "The time relations constructed by very young children are largely based on what they hear from adults, and not from their own experience... The child begins as a little relativist, but ends up a Newtonian Absolutist."

So Piaget believed that our common time notions perhaps owe more to enculturation than reality. They're learned from those precepts we imbibe from our adults during infancy. They're reinforced by those particular meanings we've been taught as attached to common time words - *past, present, future, ago,* etc.

Through our largely unthinking use of such words we then grow up to believe that 'the future' lies (somehow!) *ahead* of us, that 'the past' has (somehow!) *passed by* us, that a century is a *long* time, and so on. None of which may be obvious to the infant initially, and most of which is anyway suspect if not downright untrue...

New outlook on time

A second support for anti-memory is contained in the famous 1959 suggestion of J.L. Synge I've described in Ch.5. He stressed that time, as described by Einstein's Relativity Theory, seems largely incompatible with everyday experience.[2] So that some radical new outlook on time is now probably required

Synge suggested that this new outlook might be achieved, by giving chronometers to infants at play. Hopefully one of them might then develop some radical new outlook on time. His suggestion therefore echoed Thomas Huxley's famous advice that we must "sit down before fact like a little child" if we wish to know the truth.

In this region too lies the problem of 'childhood amnesia' – the fact that none of us have much memory of our first few years. Perhaps we were all just learning how to use past-oriented memory back then. Or having our brains "hard-wired" for it as Naom Chomski might say.

Might it therefore be that this "wiring" process was too strictly past-oriented by all that adult conditioning? If so, it might leave a possible future orientation quite latent, unrealised, ignored?

If this were indeed true, the dormant future-oriented capability (i.e. anti-memory as manifest through occasional intuitions) would also probably be subject to strong censorship. People strive towards a coherent and consistent view of reality, and so won't welcome any experience which might upset their accustomed view. They don't like *cognitive dissonance* - between what they expect to happen and then what they actually do see to occur.

If anti-memory is then really possible (as all the evidence from intuition so far indicates) one might expect it to be a highly *repressed* or self-censored faculty.

In Freudian terms repression involves thoughts unacceptable to the dominant Ego. So they're firmly pushed back, almost automatically, into the subconscious as soon as they emerge. Whence they can again only surface in moments of special relaxation, when the Censor is off-guard or asleep. Freud was also quite specific on the subconscious where the temporal is concerned:[3]

> "The processes (of the unconscious) are timeless; they are not ordered temporally, are not altered by the passage of time, in fact bear no relation to time at all".

In this conclusion he is also supported, by his one-time colleague C.G. Jung, who went further: [4]

> "We are finally compelled to assume, that there is in the subconscious, something like an a priori knowledge, which lacks any causal basis"

Elsewhere, as again seems worth repeating from Part One, Freud analysed unusual experience. He concluded that events which strike us as 'uncanny' (like coincidence or intuition) relate to the return of repressed infantile material into consciousness:

"(They involve) a class of terrifying which leads back to something long known to us, once very familiar".

In similar vein American psychologist Eric Berne (1910-70) improved on Freud in certain aspects.[5] As the originator of a new branch of psychology called Transactional Analysis, Berne described human personality in terms of 3 sectors. Of these the *Child* is the source of creative and playful intuition, best found in young people with enthusiastic attitudes.

Conversely the *Child* should flourish least in the psychologically old, those more censorious people in whom *Parent* attitudes prevail. Which can of course also explain why some people never seem to experience any intuitions at all!

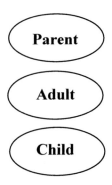

DIAG. 11.1: Berne's 3 levels of personality. The *Child* is enthusiastic, spontaneous, intuitive; the *Parent* is more inhibited, gloomy, negative. And the *Adult* tries to mediate between the two.

Finally it helps to consider anti-memory in terms of our two brain hemispheres. According to Roger Sperry (Nobel Prizewinner in 1981) our right brain sector - the seat of feeling, emotion, creativity, intuition - is primitive and relatively timeless.[6]

In contrast our left brain is more literate and logical, more conscious of time through order and sequence. So that one of its tasks is to mediate our primitive and disordered right brain feelings, into more ordered spoken words.

In our modern scientific western culture, the large emphasis on rationality tends to encourage intellect and left-brain supremacy. So we may find right-brain intuitions to be unsettling, disconcerting, even vaguely troubling. They hint at something the left brain can't really un-

CONCLUSIONS CONFIRMED

derstand. And since it appears too problematical to examine their implications, they tend to be brushed aside as of little consequence.

All of which also complements Einstein's plea in favour of intuition, as I've earlier quoted to begin Part One:

> "The really valuable thing is intuition. The intuitive mind is a sacred gift, and the rational mind a faithful servant. We have created a society that honours the servant, and has forgotten the gift"

Though here one may note that the great physicist never seems to have examined intuition very deeply otherwise. This is an important matter to which I'll return in Part Four.

In any case we can now update Einstein into more modern terms. He would then be saying, that the predominant western emphasis on left-brain logic and rationality, has degraded or obscured our right-brain intuition capability.

Conversely less literate or advanced cultures may more favour right-brain thoughts. These are not so smothered by logic, language and mathematical ideas. Such people may use right-brain or 'sixth sense' intuition, as a matter of routine.

Overall therefore various conventional contributions can help build up a more detailed and coherent picture of anti-memory capability. It's a latent, neglected and highly censored faculty. It's usually *repressed* – i.e. censored unawares - because it threatens that sense of equanimity which the conscious Ego has attained, about its relation to the outside world.

More specifically too this repression can be ascribed to conflict with our learned sense of the temporal. For unavoidably it threatens what people have learned about time - and all that they like to think they know about the same.

For me the next problem was then whether I could overcome this strong temporal censorship, to bring anti-memory under more deliberate control. And specifically in numerate situations, where hard figures could demonstrate the whole effect quite unequivocally.

It's therefore with such learned numerate abilities, through which the anti-memory effect can be developed in a more scientifically convincing manner, that Part Three is concerned....

The *most scientific* paradigm [7]

Meantime in any case it seems also inescapable, that anti-memory must be the *most scientific* paradigm for psi so far. This is because it concurs very closely with the general methodology of science in at least ten points, whereas traditional paradigms like 'ESP' or 'telepathy' concur in few or none.

First therefore in science the most basic of all procedures is to *observe the facts* with proper accuracy. However such has never really happened, with intuition research before. For people have always rather curiously failed to notice, and so never observed, that basic fact of later confirmation, which is inevitably common to intuitions of all sorts!

This new realisation of later confirmation then enables a second science aspiration. This is the *knowledge compression* of natural diversity, down to one universal summary or single common plan:

> Intuitions are always strange thoughts, apparently confirmed by observation afterwards.

As I've also just shown throughout Part Two, some 97%, of all intuition anecdotes ever published, conform to this single common plan. Which therefore enables a third science ideal – the *testable prediction* that nearly all new intuitions reported henceforth should conform likewise.

Two further aspects of scientific method are the requirement for *accurate fact-labels,* which may also *explain.* Both these requirements are well met by the two new terms 'anti-memory' and 'pre-call', as I've already considered to some depth in Ch.4.

A sixth common science procedure is the use of *diagrams* to summarise, simplify, render more intelligible. Throughout science history diagrams have greatly aided progress, by enabling an extra element of visual intelligence to come into play. And in the special case of intuitions, that simple pre-call diagram of Ch.4 more than adequately fulfils this role.

A seventh science ideal is *quantification* of whatever natural phenomenon one is dealing with. Here too the new concept of pre-call enables a reasonable first quantification of intuition frequency – at around 10^{-7} as compared with the more familiar re-call capability.

New scientific paradigms should also exhibit *harmony* with existing science in so far as possible. Such harmony has always eluded all previous paranormal paradigms. Yet harmony with existent science is readily attainable in the case of anti-memory, a matter I'll discuss more suitably in Part Four.

A ninth science ideal is the aspiration to *elegance* through concision and symmetry. This is an aspect already apparent to some degree, in those several similarities between memory and anti-memory, that I've already clarified. But in addition such symmetry can also be greatly extended, in the context of a psycho-physical isomorph with Einstein's Theory of Relativity. This is again a matter I will clarify in Part Four.

Finally in science methodology it's absolutely vital that unchecked assumptions, and further new paradigms, be subject to *validation through experiment*. Again these are issues wherein the anti-memory concept is demonstrably more productive than any other advanced so far. For it can generate new - and previously unthinkable – experiments which support the whole idea. On some of their content I'll once more be reporting in Parts Three and Four.

Objectively therefore, and regardless of subjective feeling or even unrealised censorship, anti-memory must be the *most scientific* paradigm, proposed for psi so far.

A brief summary
97% of all intuition anecdotes, as published in the literature of the paranormal, can be interpreted more simply in just anti-memory terms. This finding agrees very closely with my own two personal Intuition Surveys. In fact it agrees so closely as to suggest that there's little chance contamination, where most apparent intuitions are concerned.

Though almost totally mistaken of interpretation, the literature of the paranormal is still valuable for its personal reports from recognised intuitors. In sum these suggest that intuition is a highly repressed faculty, one emanating from the Child sector of our personalities.

Such repression then makes further sense, in terms of conflict with our accepted view of time. And as such it supports the wider concept of anti-memory yet again. All of which further suggests, that this is the *most scientific* paradigm yet proposed for psi...

PART THREE

LEARNING TO PRE-CALL

*"I sought for the barrier,
which divides our knowledge of the future
from our knowledge of the past.
And the odd thing was, I found that
no such barrier exists at all"*

J.W.Dunne – *An Experiment With Time*- 1927

12

HOW TO 'GUESS' CARDS – CORRECTLY!

"I should like to be able to perform, or at least observe, such an experiment myself"
J.E.Alcock - *Parapsychology – Science or Magic?* - 1981

NOW THAT INTUITIONS of all sorts seemed best interpreted as instances of anti-memory, my next step was quite clear. Could this repressed faculty ever be developed – i.e. improved by practice – in more concrete and numerate ways?

For, as I've earlier described, I'd observed that everyday intuitions seem to correlate with a psi-state, a fairly recognisable attitude of mind. In addition their frequency could be markedly increased, by general measures aimed at maximising this intuition favourable state.

The challenge was therefore whether I could now bring this psi-state under more deliberate control?

As yet there are no agreed words or language, to describe or communicate this rather special attitude of mind. However it can be circumscribed in its essence by various adjectives - like 'calm, happy, relaxed, cool'.

These all entail a general absence of irritations or intrusions of any sort. Or in terms more psychological, the intuitive psi-state is a 'low noise' one.

Conversely there are various 'high noise' mind-states, and these are very much the opposite. 'Rushed, harried, annoyed, excited' are some of the adjectives which circumscribe these. And in some modern lifestyles they are all too frequent - which must make it rather unlikely that intuitions would occur too often there.

Or even if intuitions were to manifest in such circumstances, it seems even more unlikely that they'd be noticed or observed!

Like many another before me, I'd further observed that strong personal experience of intuitions, soon makes the psi-state much easier to recognise. After which it grows ever less difficult to reproduce, and more or less at will. Eventually too this process of improvement can result, in more frequent and spectacular manifestations of intuition, as I've described in Chapter 2.

My challenge now was to make this general process more con-trolled and quantified. In my first two Chapters, I'd only used crude grading estimates of probability. Now I required some situation where intuition might still function, but with the odds-against-chance more firmly reckoned mathematically.

Once again too, and perhaps surprisingly, no psi investigator - nor indeed any scientist of any sort - ever seems to have gone on record with such an exercise, before I first reported my early results to J.B.Rhine in 1966.[1] I felt therefore quite unique - and truth to tell also timid - in venturing out on this new and unprecedented intellectual exercise....

Card routines

At first I tried Zener cards, as used by J. B. Rhine in his 'ESP' experiments (Ch.7). But soon I discarded these as far too dull and boring. They lacked that vital element of interest apparent with intuitions of all sorts. And in any case they were obviously unsuited to any sophisticated learning scheme.

Finally I finally settled on ordinary playing cards, very much as the first science methodologist, Francis Bacon, had suggested back at the start of the seventeenth century. He thought that the 'binding of thought' might be proven by card-guessing even if *"it hit for the most part, though not always"* [2]

But neither paranormal anti-memory, nor indeed its more normal opposite, are usually correct in full detail. So I further needed some routine where *partial accuracy* could be properly described. For if I could get anti-memory or pre-call to function, I might at first only pick up on *at least some* card details...

I started with 2 ordinary packs of playing cards combined. With these I first cast out pictures of all sorts, so that just 80 cards remained. After shuffling these I proceeded to 'guess' through them, as described more fully at the end of this Chapter.

I did this in 5 standard blocks of 5 trials apiece, a total of 25 attempts in all. After which I would re-shuffle the reconstituted pack of 80, so starting all over again. Before each 'guess' I deliberately tried to pre-call, aiming to dig out from my own future experience the details of whatever card I would next see.

Negative people, or those of no experience but strong assumptions, might well of course think that such a feat must be impossible. More pragmatically however, a little informed practice (again as I've detailed below) will likely soon indicate otherwise!

In any case I first wrote down my selection, cut the deck once and reformed its two sections, then turning over that card which was left topmost. Finally I recorded whatever outcome this turned out to be!

The cutting procedure was meant to ensure that I was really dealing with a truly future event each time. *Clairvoyance* could hardly be postulated without a great increase in imaginative extravagance. *Telepathy* was impossible because there was no other person around.

With each fresh effort I was therefore trying to pre-call, or think forward to what I might see next, about 5 seconds into future time.

I never attempted to *pre-call* the full nature (i.e. number and suit) of any card. For a mere amateur such as I was, that would have seemed too difficult by far. It would resemble those common tests supposedly sometimes given to detectives, wherein they are asked to *re-call* the full details, of some particular criminal scene.

So instead I just settled on the simplest possible 'guessing' scheme. This meant a binary system where I would select between 2 possible outcomes. For example, a red-or-black result.

Playing cards further provide three such binary modes. The next card can be red-or-black, odd-or-even, high-or-low, at all times. To get closer to the actuality of intuition as previously experienced, I therefore combined these 3 binary systems into a novel *triple-binary* form. For each single 'guessing' attempt I would then be making three binary tries instead of merely one.

For example I might choose "Red-odd-low" for my next pre-call statement. And for this 'guess' there could be only 1 of 4 possible scores, depending on the next card as yet unturned. This is a matter I've clarified in the table below.

Playing cards of course have more "High-Even" combinations (i.e. **6,** 7, **8,** 9, **10**) than "Low-Even" ones (i.e. 1, **2,** 3, **4,** 5). But in practice this doesn't matter where scores are just summated as below.

In any case my new triple-binary routine soon proved well tailored to learning in 3 ways. Firstly it allowed for *partially correct* results, which was highly desirable for those reasons I've described above. Secondly *fully correct,* and also *fully incorrect,* scores were naturally emphasised whenever they transpired. (This would be around 1 'guess' in 8 on average at first.)

Thirdly the whole procedure is three times faster than mere single-binary (e.g. red-or-black?) routines.

Pre-call		Outcome		Score/3
Red-odd-low	R-O-L	Club-8	B-E-H	0
"	R-O-L	Diamond-8	R-E-H	1
"	R-O-L	Heart-2	R-E-L	2
"	R-O-L	Diamond-3	R-O-L	3

TABLE 12.1: Triple-binary 'guessing' exemplified.

Trial-and-error always

Otherwise my overall learning strategy was just a prolonged trial-and-error routine. With each individual effort I tried very hard to relax and concentrate, very much as Mary Craig (p.67) had advised back around 1930. Always I aimed to reproduce that special mentality or psi-state, wherein I already well knew that intuitions are most likely to occur.

Each of these singular attempts was also a deliberate exercise in *failure avoidance* - attempted suppression of those mind-states usually observed to be wrong. Logically also this second aspect was even more important than the first. This is a matter I will next explain.

Suppose for example that you've just tried very hard to attain the desired mind-state – and then get a fully correct score. Still you can have no real basis for a firm conclusion in this particular case. It might equally have been chance or skill that decided your choice.

Conversely suppose you've again tried very hard to concentrate - only to get a fully wrong score! In this latter case there can be no ambiguity at all. For whatever mind-state you had attained on this occasion was quite simply wrong!

Attempted avoidance of clearly erroneous negative mind-states, rather than deliberate emphasis on less certain positive ones, is therefore the basis for pre-call learning of all sorts.

As such every single pre-call attempt is also a very deliberate exercise in *phenomenology*.[3] Earlier I had used this process with large success to attain more and better intuitions, as described in Ch. 2. Now I was aiming to narrow down the same procedures, into the more specific case of restricted card routines.

One aspect of phenomenology involves close examination and description of internal mind states. And my focus here was that elusive psi-state I was now attempting to control.

My entire learning exercise was then just one of extended and well informed trial-and-error routines. It was based on repeated and painstaking analysis of each singular attempt.

In addition I tried to transfer as much as I could from everyday experience, to help in my new pre-call learning enterprise. For example at first I conducted each learning session as a wholly private exercise. For normally when you set out to master some delicate new skill, you're likely to feel more at ease on your own at first!

Or as most amateur motorists doing the Driving Test can testify, some staring observer is very likely to disrupt competence when you're only at an early stage!

I further decided that at least 10 hours, or say 1,000 singular attempts, would be required, before any sensible judgement could be formed. Because that is the order of effort required to gain minimum competence, in many more ordinary learning feats.

My psychological input would then be informal or Shakespearean, rather than formal or Skinnerian. For Shakespeare could summarise more advanced psychology into a single sentence, than a behaviourist like B.F. Skinner (1904-90) might in a lengthy treatise.[4] And the great dramatist would probably communicate his message more effectively as well...

Each of my early pre-call sessions also involved exactly 25 triple-binary attempts as I've already explained. These constituted a 'session' split into 5 rounds of 5 trials apiece, the whole thing taking about 15 minutes in all.

For my first learning attempt I also instituted 10 daily sessions divided between 2 study periods with 5-minute intervals. Each day therefore gave a total of 250 triple-binary attempts over about 3 hours - with a best possible score of 750 and an average expectancy of 375.

'Guessing' at this rate was of course almost over-work. Still I kept it up for a period of 30 consecutive days in this first learning experiment. The challenge was novel and exciting, and I was indeed hugely curious to find out whether deliberate card pre-call might be learned!

Very quickly then it soon grew apparent that learning should indeed be feasible. So much was clear from those many secondary observations I was now amassing in support.

For example at first there seemed to be somewhat too many slightly over-long winning streaks. But these would then be too rapidly countered or neutralised, by a converse losing run. The former seemed to correspond to those lucky or 'hot' winning streaks often reported from gambling. The latter made sense as a form of subconsciously inspired catharsis – to release those tensions inspired by being too high above the expected norm.

But through an increasing sense of self-awareness, and more importantly strong self-control, I soon learned to curb these losing streaks at an ever earlier stage. My scores then began to climb away from chance quite rapidly.

By Day 30 therefore, or after some 90 hours of learning, my daily totals - around 525/750 observed where 375 was the expected average - were quite impossible by chance alone. Clearly deliberate anti-memory or pre-call was now operating on a routine basis here.

For cards therefore, in this mode at least, I could now "see into the future" by about 5 seconds – and more or less most times I tried!

The astronomical odds against chance, in this first learning exercise, could also be calculated in several ways. But broadly if anyone were to go through one daily routine of this kind, chance would never yield similar totals in a thousand years!

Extra-chance odds of a similar order also apply to those other pre-call routines I will report in the chapters just ahead. But for these there's no need to consider the possibility of chance again.

In presenting these totals for this first pre-call experiment, I've also omitted reporting on alternate days. This is because they involved rather different pre-call routines. (New ventures in research are seldom a perfectly neat affair at first!) For these, and other quite different pre-call learning modes, I gradually settled into a less demanding and more regular routine, a regular one hour per day.

Applying pre-call at this rate, in a growing variety of ever more ambitious routines, then gradually became a unique personal intellectual challenge for me. That regular daily hour, which I still allocate to various modes of anti-memory or pre-call, is more of a novel intellectual pastime than anything more serious. It continues to throw up new challenges for resolution, mostly too specialised for description at this point.

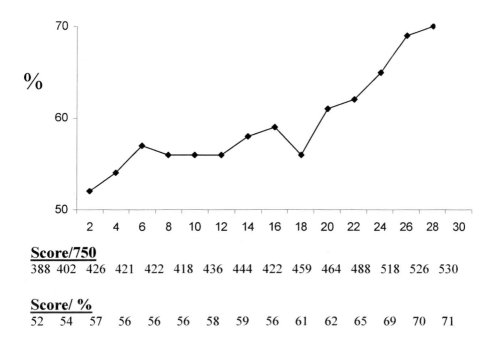

DIAG. 12.2: Graph/table summarises pre-call learning in triple-binary format with playing cards over 30 days, with totals for alternate days.

One daily hour – or more like five hours weekly - is also considerably less effort, than the time many people spend in more ordinary or familiar pastimes. For example there's bridge, golf, crosswords, SuDoku - or even just staring blankly at TV! All these are activities on which many devotees habitually spend far more hours annually, and towards less interesting ends.

In thus restricting my time to just one hour daily, I was also being careful not to let my unique pastime grow overly obsessive, or too much of an attention-demanding routine. As indeed might otherwise easily occur.

Still 5 hours weekly add up to 250 hours yearly, or 2,500 hours in a decade. And there are several decades involved in my impressions here. Which, coupled with due knowledge of the lessons of 'hard' science, perhaps affords a rather unique blend of expertise.

But at this point too I should emphasise that none of these totals I've given – nor others similar which I'll be discussing later – should be taken to imply any special or great pre-call competence on my part. In fact the converse is more likely to be true. For I've never found much reason to think that I was especially gifted at this or any other anti-memory routine.

Instead I've always regarded myself as more on the level of a rather middling golfer at any local club. One who well knows his own deficiencies, and so realises that there should be a great many others, far more proficient at the skill.

Finally neither are these results, nor those others I'll be reporting later, presented as any sort of definitive experiment. Instead they're meant to provide a sensible model of expectation, one that almost anyone should be able to emulate and indeed surpass. This they can try through those specific learning routines I've included at the end of this chapter.

All interested people are therefore invited to "Try it and see!" at their own convenience, if desired. This is by far the most common mode of proof in science for the past 400 years. It's a mode of proof which one can term Galilean, after a famous occurrence at Bologna on April 24-5, 1610: [5]

> Soon after Galileo first set up his newly invented telescope, he reported he'd seen 4 moons circling around Jupiter. However this would contradict what Aristotle had stated, and so what all those philosophers who followed him believed.
> Galileo therefore invited these Aristotelians to "Try it and see!" for themselves, by looking through his telescope. But two of them - Cremonini and Libri - refused.
> For their basic belief that Aristotle was omniscient, had already convinced them that Jupiter couldn't have any moons at all!

After 1660 a Galilean approach to proof was also adopted by Britain's new Royal Society, which speedily evolved into the world's premier scientific gathering. Its motto was *Nullius in Verbo,* which translates as *"Never believe in what you hear."*

In principle therefore, no scientist ever since believes in tales (be they tall or otherwise) reported from elsewhere. To be worth considering, such reports must tell you how to do what those others have claimed to do. Then you can "Try it and See!" for yourself if desired, so forming your own conclusions thereby.

I first published my new pre-call learning routines in 1973. [6] As far as I know however, nobody ever since has ever gone on record, as having tried out these routines for himself, Though indeed some 20 mechanised attempts at psi learning were also summarised by parapsychologist Charles Tart, in 1975.[7]

All Tart's reports were construed in terms of 'learning ESP' (sic) – but can be interpreted more economically in pre-call or anti-memory terms. Their main weaknesses were that they involved mostly children or students of uncertain motivation – and that they were never persisted in for reasons which remain unclear. [8]

There are two kinds of average
In any event successful psi learning also requires a rethink of what most people term "the laws of average". Historically these started to firm up after French scientist Blaise Pascal began corresponding with fellow mathematician Pierre de Fermat around 1654. Through the "law of large numbers" they could then explain objective chance outcomes. For example, why you will find a 50% average of red results when a suitably long series of well shuffled playing cards is turned out.

Similar reasoning applies to tossing coins or spinning roulette wheels (ignoring zero), and then became widely known as 'the laws of average'.

Note however that only *objective* matters - like the toss of a coin, the turn of a card, or the next spin of a roulette wheel – have been involved in these laws of average so far.

But a huge assumption, one generally unrealised and so totally untested, then entered at this point. People simply assumed that *subjective* guessing totals – e.g. when some guesser is either right-or-wrong in binary terms - must naturally obey these same "laws of average".

In fact however the two kinds of average are really rather different, as even some slight personal experience of informed 'guessing' will likely soon reveal. This slight experience can be quantified as about 10 hours' practice - or 1,000 individual attempts - at minimum. By then even a complete beginner should be able to appreciate those slight, subtle, differences, between *subjective* and *objective* average.

After this early stage also one should be able to observe clear correlations, between positive attitudes and correct 'guessing' results. Likewise it should be even easier to notice a contrasting correlation - between negative attitudes and incorrect results. Mixed up together, these two conflicting attitudes may well give early totals, which may seem very much like those *objective* "laws of average".

To reproduce at will that positive attitude, which generally gives correct 'guessing' results, will however take more time. It's one thing to know, but quite another thing to do! Still by deliberately striving to maximise this mind state and minimise its opposite, learning can very naturally proceed.

About the only previous effort to similar depth, was that conducted by some of J. B. Rhine's earliest card-guessing students during his vain quest for ESP (Ch.7). But their *individual* efforts were never at all so extensive, as those I'm now reporting here.

This particular sector, of *subjective* chance and average, has therefore always been most jumbled and confused. In fact it has caused renowned biologist Sir Alister Hardy to query: [9]

> "Is there something about probability which we don't understand?"

To which a clear answer can now be given unequivocally. *Objective* and *subjective* averages are not at all equivalent, as previously assumed. Instead there's a slight and subtle difference between the two. This difference makes initial learning feasible, so that eventually it becomes obvious and inescapable.

A new skill of mind

Soon after Day 30 in any case, I abandoned triple-binary 'guessing' for the moment, to investigate other, more sophisticated, pre-call routines. They were applications of much greater interest than mere binary pre-call could ever be, and over the next few chapters I'll be describing some of them.

Much later also I was able to reproduce successful card-guessing in computer mode, though never at all to such high levels as with more everyday routines. My lesser success with computers may be due to some unrealised personal idiosyncrasy. More likely however the machines bring in several novel factors, which may militate against the delicate pre-call faculty.

For example there seems to be something, about the actual task of pressing numbers on a keyboard, which can greatly degrade one's usual pre-call competence. In addition computerisation must involve other complexities, issues not properly explored.

One such is the totally unchecked presumption, that sub-microscopic state alterations inside some silicon chip, are in essence similar to everyday experience. Though the latter of course are on a much larger scale.

Then there's the timing process that decides which random numbers will appear on screen. For example C. Tart (as above) reported on two electronic guessing machines, whose outcomes relied on exactly when (to a matter of micro-seconds (10^{-6} sec.)) the guesser had pushed his chosen button.

Temporal discrimination to this level is of course thousands of times beyond human competence, and is in any case wholly absent from intuitions in their natural milieu. So that at the very least a new, needless and totally unexamined complexity has been introduced.

Card 'guessing' therefore remains the most cautiously sensible medium for pre-call development at an early stage. It's best to stay 'low-tech' but sure in your learning, at least until some minimum experience and competence is gained!

Meantime I now had some dozen firm conclusions, about learning to pre-call in a numerate way:-

1/ One can attain rudimentary success with card "guessing" - to 70% levels in a 50% chance situation – through deliberate employment of pre-call concepts, during ca. 100 learning hours.

2/ But 70% scoring means only 40% competence or skill! Suppose for example that in 100 trials you make just 40 skilled or correct ones. You will then be left with a further 60 unskilled trials, from which sheer chance will afford around 30 correct returns. Add this unskilled score to your previous skilled one, and you will have 30 + 40 = 70% scoring overall!

3/ Learning could likely be carried to higher levels of competence, if further pursued.

4/ Test variations to the usual routine made no real difference to my competence. For example I could still pre-call at the usual rates *without* cutting afterwards – which would have been judged as *clairvoyance* in those old outmoded terms!

5/ There was often a strong secondary effect, of too-close adherence to the current mean. For example over Days 6-18, each total differed very little from the current (427) average.

This effect resembles the common experience of "current form" - as found in other high-concentration activities like golf, darts, marksmanship. It supports my general impression, of deliberate pre-call as a similar high-concentration skill.

6/ The learning process seems more a minimisation of emotive abreactions, rather than attainment of some completely new skill. As indeed might be expected, from a natural but repressed faculty.

7/ The sum of my findings also supported my growing under-standing of anti-memory, as a latent and undeveloped capability. It's easy to recognise but extremely difficult to communicate. - primarily because we lack adequate language for doing so.

8/ My feat of paranormal learning was relatively easy – but only because I'd gone about it with a clear rationale never applied before. For throughout the whole failed saga of paranormal research (at least 1 million man-hours in total), nobody ever seems to have attempted learning via such routines.

9/ A total beginner might hardly expect to master numerate pre-call as rapidly as I did here. This is because I had much experience of intuitions in general for several years before.

10/ But otherwise there seems no reason why most sensible people shouldn't be able learn likewise. For - apart from its major novelty of time inversion - there seemed little mysterious, 'psychic', or otherwise paranormal, about pre-call skill.

11/ This conclusion is supported by various others, who at different periods attempted learning for a while. Mostly these learned from initial exposition by me, and then from those several modes of instruction I have provided here.

12/ All rightful critics, doubters, disbelievers – like J.E. Alcock whom I've quoted at the top of this Chapter - can therefore now be invited to "Try it and See!" in their own experiments.

Meantime my original concept of anti-memory had emerged from card 'guessing' much strengthened once again. Instead of being just a mere hypothesis, it was now tending towards the higher status of theory supportable by experiment. The next question was therefore whether further experiments might support this status still more….

Learning Instructions: How to pre-call cards

Take 2 new packs of ordinary playing cards. Cast out all pictures to leave you with a total 80 cards. Leave these in a shuffled stack, with backs of course uppermost! Always cut and re-form before inspecting the topmost card.

Set such inspected cards aside (face down!) afterwards, until you've dealt with 25 attempts from the original stack. Then recombine all cards and shuffle thoroughly.

Make each individual 'guess' or forecast in triple-binary form initially, as I've already described. In other words, chose simultaneously between the three option possibilities: high-or-low? odd –or-even? red-or-black?

So your choice might be something like 'high-even-red'. (Low cards are numbered 1 through 5, high ones 6 through 10.)

Now try to follow the following instructions as best you can:

1/ First make some attempt to 'steady the mind' and create that calm, cool, relaxed mental attitude which pre-call requires. After taking about 15 seconds to do this, record your choice in abbreviated form. For example r-o-h for red-odd-high.

2/ Next cut the deck once, re-form it, and remove the top card. Place it well apart and upturned so that you can see what the 'answer' or outcome is. Record its details in larger letters – e.g. B-O-H if it was black-odd-high – alongside your original forecast. Then lay this card aside, but reversed again to stop its details intruding on your gaze.

3/ Record also your score – e.g. 2 out of a possible 3 in the example above. Reflect briefly on this score and try to describe to yourself your mind-state as your choice was made. For example did the correct (i.e. odd-high) elements seem clearer than your incorrect (i.e. red) choice?.

4/ But if your attempt was fully correct (e.g. 'guessing' r-e-h for an R-E-H outcome – max. score of 3) then pause for a more detailed analysis. What was your mind-state when you made this perfect try?

5/ In such cases proceed *quickly* to your next attempt. It may be that you've now attained the desired psi-state, and it usually persists for a while.

6/ Conversely if your attempt was fully incorrect (e.g. 'guess' b-e-l for a R-O-H outcome), pause for an even longer analysis than before. Were you somehow distracted, instead of concentrating as you meant to do?

Take time to restore your cool mental attitude in the desired direction, before proceeding *slowly* to the next attempt.

7/ Have a rest and a brief review after every 5 attempts, regarding this group as a mini-series on its own. Was your score near the maximum (15/15), or conversely closer to the minimum (0/15)? Can you see any correlations between your attitudes and such scores?

8/ Before starting each mini-series, aim only *not to lose* (i.e. score less than 8/15) at first. Later you can gradually increase your target (to 9/15, 10/15 etc.) as you gain in experience and confidence.

9/ Likewise conduct a longer inquest and analysis after each 25 trials, or a total of 5 mini-series in all. Then aim to deliberately raise your scoring away from its "chance average" (i.e. 37.5/75), at a suitably gradual and easy rate.

10/ Each series of 25 trials conducted in this way should take about 15 minutes. You should aim for 4 such series – or 100 triple-binary trials – every day. 50 daily trials might be sufficient; 200 trials daily would be too much.

11/ Record your overall totals neatly in a separate record of easily assimilated form. Aim to increase your daily average scoring (around 150/300 at first) by just 1 or 2 each day. Maintain a 5-hour rolling total for each working week.

After 10 to 20 *regular* practice days or hours (spread over 2 to 4 weeks) you should then be aiming for an average around 800/1,500 – i.e. 53.3%. *at least*.

12/ As with other more normal high-concentration exercises like golf, darts, marksmanship, you must keep up regular practice every week. If not, your hard-won pre-call competence will probably degrade quite rapidly.

13/ At first beware especially of severe down-surges after particularly good scoring interludes. Learn to see the former as the obscuring and emotive abreactions that they really are.

Remember that your internal Censor won't really like to see your pre-call ability increase, because this must inevitably confuse its previously complacent time ideas.

14/ Conduct all sessions *in private*, with distractions minimised at first. Only later should you consider demonstrating to others or in public – i.e. after sufficient private competence has been attained.

15/ Continue in this way with no major training lapses for 10 weeks or 50 practice hours. Realistically you can aim to attain reliable 70% competence by then….

A brief summary

To learn numerate pre-call – in triple-binary routine with ordinary playing-cards – proved an unexpectedly simple exercise. It's largely based on informed trial-and-error, over a sufficiently long training period.

In the 50% binary situation, routine 75% scoring proved not too difficult to attain. At this competence level many useful insights become readily available. One is that similar pre-call competence should be equally attainable, with any sensible future-oriented exercise.

Those reports and learning rules, which I've furnished, should further make this experiment quite repeatable. They should likely enable just about anyone, to "Try it and See!" at their own convenience as desired.

In this way longstanding doubts, about whether paranormal experience is a valid aspect of reality, can easily be brought to an end. Surprisingly, no scientist ever seems to have reported any similar learning attempt, to the required depth, previously. But through such my original hypothesis of pre-call or anti-memory is again much strengthened, and to a higher level than before.

13

3-DIGIT PRE-CALL

"Science is merely the search for the one in the many, or more exactly the search for unity in the wide variety of personal experience"
J.Bronowski – *The Ascent of Man* - 1974.

BUT WHEN ALL is said and done however, pre-call in binary mode is still just about the most limited form of competence imaginable. So I next searched around for a second and wider test for the anti-memory concept. Ideally this would involve routines more sophisticated and informative, while still strictly numerate.

I found such a system in 3-digit pre-call, a procedure again suggested by that ever-useful example past-oriented memory. In his book *About Memory* (1885), early psychologist H. Ebbinghaus has then reported a famous self-experiment, one that has since proved largely reproducible.

In short he committed himself to learning off long lists of nonsense syllables – like XOY, GRH – in a fully private routine. Then he checked on how well he could recall them after varying intervals

In this way Ebbinghaus formalised the intuitive idea of a *memory curve,* which I've already used in Chapter 4. A memory curve diagrams the common experience that all our memories tend to fade away with time. His findings are now largely accepted by psychologists. Though in truth not too many of them ever seem to have repeated, that original experiment as reported by him.

These Ebbinghaus experiments also led to *digit-span* tests for measuring normal re-call capability. Here people are shown a glimpse of some very long telephone number, for example - 352584369276. Then they are tested on how much of it they can re-call after a set interval.

And whereas 9 to 12 digits can usually be reproduced by alert young students, 3 digits might tax very old people with poor re-call capability.

Since the proposed paranormal faculty seemed much weaker than the conventional or normal one, a 3-digit span also seemed to me about right, for the converse testing of anti-memory. In normal memory testing one aims to *re-call* the relevant digits *after* they've been observed. But here I would aim to *pre-call* those relevant 3 digits *before* they were actualised!

To generate 3-digit future outcomes, I first used figures from a random number book.[1] Such books just gives endless lists numbers all mixed up.

But later, for continuity, I switched back to ordinary playing cards. This merely required that I would try to pre-call whatever 3-digit number, would be produced by the next 3 cards successively turned up. (10 must be read as 0 in this exercise.)

In short then I soon began to realise, that pre-call could again be made to function in this mode. Thereafter my competence again started growing by the hour, all very much as before. In addition 3-digit pre-call proved *more interesting*, and so easier to assimilate, than my earlier triple-binary mode.

Likewise again I can see no reason, why my broad pattern of learning should not be repeated, by anyone who so desires.

In any case 3-digit pre-call executed in this way, soon proved to be a most informative routine. It threw up many unexpected but illuminating lessons for me to contemplate. Mostly they reinforced the general picture of anti-memory as a highly repressed faculty of intellect. Some I'll describe at the end of this chapter below.

But the nature of correlations – between pre-call and reality as the results piled up – proved also to be a rather sophisticated one. To describe these correlations properly therefore required a special and slightly complex scoring protocol. This I will describe in the next section, though non-mathematical readers may prefer to skip lightly through it without much loss overall.

Scoring conventions

Firstly therefore all numbers must be pre-called in 3-digit form. So if you select some number like **69**, that must be recorded as **069**. Likewise if you have chosen just the figure **7**, it must be entered as **007** in your records.

The second restriction is that all pre-called digits must be different – even though this may often not be true of the result. For example you might pre-call **947** where the target turns out to be **444**.

3-DIGIT PRE-CALL

This artificial restriction on choice is an unfortunate necessity, if you wish to calculate the exact odds-against-chance at first. Later it can be dropped, when you've developed high or non-marginal pre-call competence. For at that level probability, or odds-against-chance calculations, become no longer relevant.

Pre-call	Result	*UNPLACED* Score	Chance/100	*PLACED* Score	Chance/100
7 1 4	5 3 **1**	x - -	65.7	- - -	
6 9 3	8 **9 6**	x x -	15.0	+ - -	30.0
5 9 2	**5 2 9**	x x x	0.6	+ - -	30.0
0 5 3	**0 3 3**	x x -	15.0	+ + -	02.8
0 1 9	**0 1 9**	x x x	0.6	+ + +	0.1

TABLE 13-1. Example of 3-digit scoring protocols

To describe the observed realities of 3-digit pre-call, a special scoring system is further required. You need two separate scoring columns – a *unplaced* and an *unplaced* one. In these you record each single digit success, always starting to the far left in each column. It also helps to tot up all such successes after each 5, 25, and 100 trials.

Examples of these *unplaced* and *placed* score records are given in the table above, along with their chance probabilities.

Initially of course also you can hardly expect to pre-construct your 3-digit number fully, by pre-calling all digits in their proper place. Since early pre-call is always a weak and undeveloped faculty, you'll probably score just 1 digit out of 3 initially. Later 2 digits should become more common – after which full 3-digit correspondence can gradually ensue.

But at first also your weak single-digit correspondence is rather unlikely to be sited in its proper place. More likely you'll pre-call something like **8** 3 6 for a 0 **8** 2 result. And when double-digit correspondence later commences, it's unlikely to be placed properly likewise.

More or less as with any other complex skill in real life, 3-digit learning is then a gradual process. Like learning to drive, or to typewrite, or to play the piano, it will take at least 50 hours of practice – say 5,000 individual trials - before you can classify yourself as even an amateur.

Overall too the whole 3-digit routine is a rather exact temporal inversion of Ebbinghaus, though with digits instead of alphabet letters! And it comes replete with so many informative special effects (some half-dozen of which I've noted below) as could keep whole teams of psychologists busy for many years.

How to learn 3-digit pre-call

1/ Provide yourself with a reporter's jotter, which you should rule into 5 columns about 1-inch wide. These will record 'guess', result, non-placed score, placed score, general remarks.

2/ Likewise rule every fifth line horizontally to provide 4 blocks of 5 'guesses' each per page.

3/ Take about 15 seconds to calm down and concentrate; seize on the first 3 digits that come into your mind; record these in Column 1.

4/ Cut your combined deck of 120 cards (not 80 as described in Chapter 12), and turn up the first 3 cards in order, to give your 3-digit result. (Read 10 as 0 here.) Record this result in Column 2. Then enter your scores, *unplaced* and *placed* as I've described above, in Columns 3 and 4.

5/ Scrutinise your performance carefully after each individual trial. If it was an especially good score, try to clarify that state of mind in which you chose the same. Were you especially relaxed? unusually careful? particularly happy?, concentrating hard?

6/ Conversely if your performance was an especially bad one – i.e. with no score of any sort! - conduct a similar analysis. Were you especially rushed? unusually careless? too excited? thinking of something else perhaps?

If so, pause to collect your thoughts before proceeding very carefully to your next trial.

7/ Add up your totals after every 5 and 20 trials. Conduct a general inquest at these points, resolving to do just slightly better next time.

8/ Aim always for a standard 100 trials per session, which should take just over an hour each day. After the first week of 500 trials, you should be starting to recognise, that special psi-state in which pre-call operates.

9/ Progress thereafter is largely a matter of trial-and-error routines allied to will. It should afford a gradual decrease of negative scores at first. Thereafter expect a slow transformation, of *unplaced* **x** scores, into more correct *placed* **+** results.

Many special effects

As I've mentioned earlier, 3-digit learning also threw up some half-dozen special or secondary effects. Mostly these were quite unexpected; few or none might reasonably have been predicted beforehand. Some of these were confirmed later by others, as I will relate. These effects also reinforced my growing belief, that the skill is indeed a time-inverted analogue to past-oriented memory. And further one likely to be general rather than purely personal.

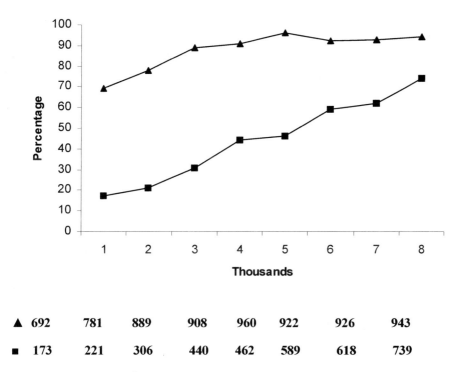

| ▲ | 692 | 781 | 889 | 908 | 960 | 922 | 926 | 943 |
| ■ | 173 | 221 | 306 | 440 | 462 | 589 | 618 | 739 |

DIAG. 13.2 : A 2nd experiment in 3-digit learning, conducted 5 years after the first. Noting only unplaced X (▲) and XX (■) scores.

Like some explorer returned from unknown territory, I'll therefore detail these special effects now. They should prove useful as 'marker beacons', for future investigators who may choose to pass this way.

Firstly therefore it became ever easier to pre-call just single-digit correspondence, and usually not placed where it should be. For example I might typically pre-call **2** 3 7 for an outcome which turned out to be 9 **2** 4. But as learning progressed I quickly began to pre-call more digits into their proper place – for example 6 **2 9** for 0 **2 5**.

Secondly, as 1-digit scoring became ever easier, better *partial* or 2-digit correlations began to increase. And at this stage such a capability - to get just 1 or 2 digits correct instead of the full 3 - made eminent sense. It's yet another paranormal similarity with normal or past-oriented memory.

For example traffic police frequently encounter a similar effect. This happens when hit-and-run victims can only re-call 1 or 2 digits of the offending license plate, less often 3 or more.

A **third** special effect became very apparent around this stage as well. This was a very marked tendency to pre-call 'chunks' of the outcome - with 2 digits in proper order, but not placed correctly overall. Or in other words I was making too many choices like **5 1** 4 for results like 7 **5 1**.

While it must seem almost impossible to assign clear mathematical odds-against-chance to this *'chunk' effect*, at this stage it appeared so prominent as to be almost unmistakable. It may well have something deep to convey about memory retrieval in either past or future mode.

In any case, as my 2-digit scoring or correspondence gradually rose towards higher 3-digit levels, a **fourth** and very remarkable effect of *contiguity* appeared. This means a most striking tendency to pre-call ever *nearer* to the actual target digits, while still not getting them totally correct.

For example in the initial learning stages, I might pre-call something like 0 9 4 for 2 6 7, i.e. with all digits hopelessly astray. But in later stages I tended to pre-call numbers like 3 **6** 8 for the same 2 **6** 7 outcome. And while there is only one digit fully correct here, the other two are *very nearly* correct as well.

Again it seems almost impossible to assign chance odds to this effect: for one thing it depends on subjective guessing habits. Still this *Contiguity Effect* was observed independently by my colleague Frank Morgan from Atlanta, Georgia when he undertook 3-digit learning – and without any prior reference to it on my part. So it seems only fitting to name it the Morgan Effect, confident that it will serve as another prominent 'marker beacon' for those repeating this exercise.[2]

All in all too this Morgan Effect seems best explained in repression terms. It's as if one's subconscious mind were still trying very hard to keep pre-call underground, refusing to let those desired fully correct digits surface into full consciousness.

A similar effect has also been reported by Charles Tart, in those psi-learning experiments I've mentioned earlier (p.129). He found that some of his best subjects too often tended to guess, *near to* the target number by plus-or-minus 1, while still not totally correct. For example if the target card or light was numbered 5, then they would chose 4 or 6 too frequently.[3]

In keeping with his dominating paradigm of ESP, Tart interprets such facts in terms of 'spatial focusing'. Still it's simpler to think in terms of common past-memory habits. For example, if you're now asked to recall a personal date from some weeks ago, you may say 'about the 13th' - when really the 14th was the actual date involved!

In any case the same factor of repression could also be read into the **fifth** or *mirror-image* effect. This became very noticeable around the advanced learning stage, where full 3-digit correspondence was starting to come through. Then I discovered a marked tendency to pre-call *mirror images* – for example **6 9 8** for an **8 9 6** result.

<u>**6 9 8 | 8 9 6**</u>

DIAG. 13.3: 3-digit numbers were pre-called as mirror-images initially.

But again of course this is just what you might expect from a nascent anti-memory faculty, one not yet accustomed to sorting out the difference between past and future perspectives. And indeed J.W. Dunne (Ch.10) had also reported similar inversion – though from left to right in his case – in his 'wild-horse' anecdote. [4]

Repression could also be read into my **sixth** and similar special effect – a marked reluctance to attain fully correct, i.e. *placed,* scores initially. Sheer chance would indicate that you should attain 1 such total correspondence in about every 1,000 trials at all times.

Still it took me over 6,000 individual trials – in an earlier series not reported here - before I got my first fully correct, i.e. *properly placed,* result. Which also happened to be the rather memorable number **4 3 2** !

But once this 'psychological barrier' had been broken, full perfect 3-digit correspondence then proved ever easier to attain. It all felt a bit like the classical mental barrier of the '4-minute mile' which Olympic runners once had to confront – a barrier ever easier to surmount once it had been overcome initially.

Finally there's a **seventh** special effect, which unlike those others seems eminently predictable. Indeed it's only to be expected from any demanding intellectual exercise. In short, as with intuition in general, one's pre-call competence tends to correlate, with just how intellectually fit you feel.

So common anti-intellectual factors – like overwork, lack of sleep, background anxieties, hangovers, etc. – will all degrade one's pre-call competence markedly. Indeed even heavy reading on serious matters can diminish capability, for maybe an hour or so afterwards. All this too is also very much as remote-viewers have reported for what they termed 'inclemencies'. (See Ch.9)

Conversely a balanced lifestyle – with good rest and a general feeling of well-being – will afford heightened levels of pre-call competence. Which again just confirms my overall impression, that it's much like any other demanding intellectual exercise.

A brief summary

The re-call experiments of H.Ebbinghaus, which used random letters a century ago, suggested that a time-inverted pre-call version might be informative. For this I used random 3-digit numbers, mostly generated by playing cards as before.

3-digit pre-call in this manner proved to be readily feasible, and also very informative. It generated some half-dozen unexpected special features, none of them very predictable otherwise. In sum they reinforced my original conclusion - i.e. that anti-memory really is just a time-inverted paranormal analogue, to the more normal or past-oriented mode.

But the 3-digit system also proved very effective to label real life events at much greater future intervals, a matter I'll leave for the moment until Ch. 19…..

14

ROULETTE AS A GAME OF SKILL

"No one can possibly win at roulette, unless he steals money from the table when the croupier isn't looking."
Albert Einstein

I NEVER ENTIRELY ABANDONED pre-call with ordinary playing cards. They provide a convenient litmus test, which can readily gauge one's level of pre-call competence at any given time. For example while writing these chapters I returned to binary pre-call with cards (Ch.12) on an hour-per-day routine. And so I rose quickly from pseudo-random 50% scoring, to regain my old levels of 70% within 2 weeks.

Though of course, as I've explained earlier, 70% scoring is never as good as might at first appear. In reality it just means 40% competence, i.e. that you're pre-calling correctly on just 40 out of 100 trials. Sheer chance will add an average extra 30 correct scores for the remaining 60 trials, so giving your total of 70% overall!

As with most things in anti-memory however, the real situation is considerably more complex than I've just shown. For one thing there's the problem of *psi-missing* - that frustrating tendency to do wrong instead of right even when you're trying very hard. *Psi-missing* was probably also J.B.Rhine's main discovery amid all that confusion about ESP (Ch.7).

However with other high concentration exercises also, one can often almost deliberately do *wrong instead of right*. Ask any golfer who has just played the first 9 holes to a record score, and watch him trying to explain why it's so much harder to keep up this level for the rest of the game.

A clearly future-oriented exercise

But in any case, and regardless of that crucial procedure of cutting the deck before inspection, card pre-call still wasn't quite totally convincing as an unarguably future-oriented exercise. To some extent that future information is always *already in existence,* at the moment you decide. It exists as a card still not turned over, even though its final destination on top of a deck may not yet be actualised.

Then too there's the traditional association between card-guessing, and those old superfluous notions of *telepathy, clairvoyance, ESP.*

What I therefore required at this stage was some equally simple, but more definitely future-oriented, exercise. It should be a mode where the future outcome was *not yet in existence,* at the time when the original pre-call choice is made.

And the obvious answer to this requirement was roulette, the game with a spinning ball and counter-rotating wheel! Some have attributed its invention to Blaise Pascal when he formulated the laws of *objective* average around 1654. Though perhaps more likely it had been known since ancient Roman times.

On the classic European roulette wheel there are 36 numbered slots along with a single zero; the much greedier American version has two zeroes instead of one. The zero affords a crucial edge or profit to those who run the wheel. In the European game this profit varies between 1.4% and 2.8%, dependent on how your bet is placed.

Originally the zero sign may even have represented $^\circ$ – the sign for degrees which was originally the Babylonian symbol for the sun. They reckoned there should be 360 days, or sunrises, in the ideal year!

To set the wheel spinning, with the ball counter-circling around it, will also give you a method of randomising far more reliable than just shuffling cards. And if you ignore the zero whenever it turns up, roulette further lends itself to 3 binary descriptions as before. In effect you can then pre-call red-or-black?, odd-or-even?, high-or-low? – all very much as with cards previously.

The real crucial difference, between cards and wheel, is of course that roulette is most definitely oriented into future time. The outcome is *not yet in existence* at the time your choice is made. So that success here must provide an even more convincing proof, that pre-call or anti-memory must really be at work.

In thinking like this, I was of course ignoring psycho-kinesis, i.e. mind-over-matter possibilities. But these are potential paranormal effects which seem to be very rare in reality. Indeed I'd only encountered them just once so far. And in any case they could be excluded more definitely in later trials (Ch.19).

I therefore approached this new challenge - i.e. trying to pre-call roulette - with large curiosity and even some trepidation at first. It would provide the first really acid test for my whole anti-memory concept. For the *future* reality was clearly not yet *present* when I would attempt to pre-call the same!

Re-learning all over again!
First therefore I provided myself with a miniature roulette wheel, one of good quality and 8 inches in diameter. On giving its rotor a vigorous flick it took about 15 seconds, for the steel ball to circle and tumble down into whatever numbered slot. A few hundred test spins soon showed that the wheel was operating properly – sufficiently true or unbiased for my ends.

Would pre-call then operate over this new and greater 15-second interval into future time? All my experience so far now indicated that it should. And yet the answer to this most crucial of all questions, seemed very far from clear at first.

This was because of my relative ignorance at this stage. Rather foolishly I had assumed that my undoubted pre-call competence elsewhere, would be immediately transferable. Or in other words I was expecting, that my current binary scoring with playing-cards – about 75% on average – should work equally with the wheel. I should therefore be able to score immediately at similar levels in this fundamentally different mode!

However this proved to be a totally misguided expectation on my part! With roulette at first, it was all very much as when I had started on cards long before. So that my early brief episodes of success, were too often followed by counter-episodes which failed. These were down-surges which wiped out my initial gain.

The result was superficially random totals over many aborted trials and abandoned days. Clearly therefore high pre-call competence in one mode, need not transfer automatically to a different routine!

Gradually then I came to realise that my initial expectation here, was in any case unreasonable. Automatic transfer won't usually happen with any more normal concentration skill. For example with music you may learn to excel on the piano – but still prove stumbling when you then start on the violin! Or of course vice-versa as the case may be.

In general therefore, when you take up any new skill after having grown proficient at a related one, the input from your first experience will prove helpful rather than complete!

After a very slow assimilation of this new clarification, I then went back to roulette as before. My basis for learning would still be trial-and-error as previously. Before each individual trial I would try very hard to reproduce that elusive psi-state which had usually proved positive. Conversely I tried even harder to eliminate other mind-states more clearly realised as negative.

To a certain extent therefore, I was in effect *re-learning* to pre-call in this new mode all over again. It was somewhat as when you sit in to drive some new car model after years in your own familiar vehicle. Then you are fitting an old and familiar routine, into a new one slightly different!

Quite slowly at first, and after maybe 50 hours involving different attitudes and mental approaches, I therefore began to pre-call roulette successfully. It was then indeed all very much as with cards before. Gradually it clarified that I could again pre-call with good reliability, over about 15 seconds into future time. This was the average interval between choice and result, as the circling ball finally settled into whatever numbered slot it chose.

Inverse Expertise?

In thus being able to predict roulette successfully, I was of course disproving, what so many eminent mathematicians and physicists had often proclaimed over the centuries. All were united in proclaiming that roulette prediction must be quite impossible Their ranks included the Bernouilli brothers, Laplace, Poisson, Poincare, Claude Shannon and J.W. Gatling of the Gatling Gun. [1] In this they were even joined by Einstein, whom I've quoted to head this chapter.

Still not even he ever seems to have investigated either intuition or roulette reality at practical first hand. So that all these authorities were just projecting *theory* from objective physics, never considering that in *practice*, subjective psychology might have a different tale to tell.

From all of which also a valuable lesson could be drawn. Large theoretical expertise for mathematics or physics, need not at all imply similar competence in other areas outside the chosen field! In fact, if anything, the correlation seems more likely to be an inverse one.

For example, inverse correlation was quite inescapable in the case of William Rowan Hamilton (1805-65), famed Irish mathematician on whom I have written a psychological biography.[2] Hamilton was the inventor of new kinds of algebra, wave-particle mathematics, quaternions for the fourth dimension, and much else besides.

But the more he immersed himself in such esoteric mentalities or thought habits, the less he became aware of normal interpersonal relations, or other more everyday concerns. As indeed the Dublin populace, and of course those best acquainted with him, well knew.

Further examples must include Lord Kelvin, the foremost British physicist of late Victorian times. He was much expert in matters thermal and electrical. And yet in his later years Kelvin opposed Darwin's theory, radioactivity, car speeds above walking pace, and powered flight. This last he regarded as impossible!

All of which may also suggest a slight new law of psychology or history, one that I'll term *"The Law of Inverse Expertise"*:

> The more effort an expert spends at specialising in one discipline, the less effective he may well be, in other very different fields

Or as J. N. M. Tyrrell expressed *Inverse Expertise* long ago: [3]

> "(A specialist) tends to see the universe, through the tinted glass of his own spectacles.... (Yet he) knows only his own branch of science thoroughly and cannot speak *ex cathedra* about others.
>
> "As regards their significance, he is in no better position to speak, than a man who is not a specialist.
>
> "He may even be in a worse position, for his special knowledge may colour his outlook."

This further has much to do with strong seeming conflicts between time as described by physics and time as experienced in everyday life – a matter best left until Part Four.

Secondary Effects

In any case it's now worth considering 3 roulette series I amassed around this time, because of the varying characteristics they reveal. These I've summed up into a single series with 3 main divisions, all of them results from my own trusty wheel. And in the next chapter I will summarise a second series, but now in a public casino milieu.

As previously with my card totals, none of these reports is presented now as some especially definitive or cast-iron experiment. I'm not writing for some *Journal of Statistics* here. So instead they're more like a useful model towards which others might aim. They're a first account, brought back by a scientific observer, reporting his exploration of new territory.

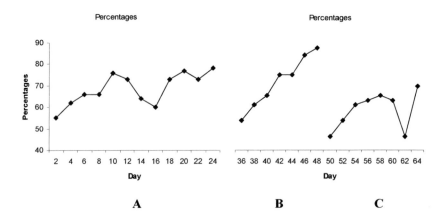

DIAG. 14.1: Results with private roulette over 63 days in three episodes, with 50% chance expectation always. Zero outcomes are ignored.

Further, as with my two earlier experiments in learning, I've marked out this third novel trail of exploration, by various secondary landmarks along the way. Few of these landmarks could have been expected by reasoned prediction alone.

And I fully expect that any followers must likewise encounter them, if ever they venture this far into anti-memory territory. Once again therefore my reports on learned roulette are best regarded, as a prototype model to which others may reasonably aspire - before improving on the same.

Part A of this series – over Days 1-23 - then shows how learning progressed with a daily average of 50 double-binary (e.g. red-high) trials. I commenced from a relatively high base around 55% because of those earlier preliminary attempts I've mentioned above. And I ended on a high scoring plateau around 80%, between Days 19-23.

Then a two-week holiday intervened. Like a golfer returning to practice on Day 36, I'd somewhat forgotten exactly how to execute the old routines. (Deliberate pre-call is an extremely labile and delicate effort of mind!) With Part B – over Days 36-48 - I had then more or less to start all over again with low pre-call competence. It required over a week to regain my previous high 80% plateau.

This *Disuse Effect,* which I observed here, is also one I would encounter repeatedly, with all other anti-memory modes. So that eventually I had to accept, that pre-call competence, in any specific mode, can decline very rapidly without regular practice. But then it can always be readily regained when practice is resumed. Further the skill is regained more readily with each successive return.

Here, as before of course also, the novel anti-memory exercise seems little different from other high-concentration skills. For example all athletes well know, that two weeks without any practice, may result in a temporary performance drop on their return.

Part C, over Days 49-63, further demonstrates that pre-call can be made to function in the presence of another person just as well. The 'croupier' here was my friend Norman, who kept his own records. Further he controlled the ball and wheel, now visible to the two of us.

Inevitably also Norman's sheer presence now altered my former studious and totally private milieu. It meant that more complex variables of psychological interaction had been introduced. But luckily he was a cheery, positive and always enthusiastic observer.

So that, even though it was more difficult than with total privacy, it wasn't too hard to keep up good scoring with Norman in control.

This series was also notable for the gross anomaly of Day 61 - which produced a total loss in competence! This shows as a sudden and violent dip in the overall tables and graph above. Here my current 'form' - around 67% by daily average - was hugely disrupted, by a wholly incompetent 46% score!

Luckily however there seemed little mystery here. My below-average scoring was clearly correlated with an over-long, and indeed over-alcoholic, party of the night before!

But again of course the same sort of thing is only to be expected with any more normal high-concentration exercise. For example you won't find professional golfers over-indulging on the night before big tournaments. My poor performance on Day 61 might therefore easily have been avoided, had I been more experienced back then.

Around this stage I also tried demonstrating anti-memory before several other singular observers not reported here. And with some of these other observers high pre-call competence proved very difficult – indeed almost impossible - to reproduce!

At first I found this *Observer Effect* (one quite notorious in other studies of the paranormal) to be yet another highly puzzling effect. It seemed almost impossible to communicate or explain. But later I realised that golfers, chess-players, or indeed most athletes, often experience something very similar. In hostile atmospheres they find it more difficult to perform well.

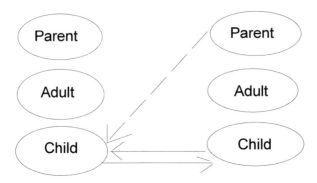

PRE-CALL ADEPT   OBSERVER

DIAG. 14.2: Childlike intuition can demonstrate readily under complementary transactions, or parallel lines. But the Child may refuse to function if Observer's stroke is more Parental or negative (diagonal line).
In more everyday experience *crossed* transactions like this, can often lead to communication breakdown.

Gradually therefore, this initially problematic *Observer Effect* clarified as before. It's just a predictable subtlety of interpersonal transactions as formulated by psychologist Eric Berne, a matter Diagram 14.2 explains.[4]

And with these various secondary effects therefore clarified, I'll next provide new roulette learning instructions, which are all very much as with cards before.

How to pre-call roulette!

Before tackling roulette you should first amass a minimum 20 hours pre-call practice with playing cards as already described. This will afford you some minimum experience in recognising the relevant psi-state, and also its opposites. These you will now seek to control in the somewhat more challenging roulette mode.

Next provide yourself with a good quality roulette wheel at least 6 inches in diameter. These are available in most good toy or games shops for about £20 or less. Discard those rakes, plastic chips, and felt cloths, which usually come with them. You only require the ball and wheel -

where a vigorous flick will afford a random result when the ball finally settles about 15 seconds afterwards.

If Zero comes up – which it will at an average 1 time in 37 on a European wheel – record it with a special tick alongside your scoring columns. But ignore it otherwise. This procedure simplifies scoring overall. Repeat that pre-call choice for which Zero turned up, applying it to the next result.

Roulette can be scored in exactly the same binary terms, as I've given for cards in Ch. 12. Though here 'low' numbers will mean 1 through 18, 'high' numbers 19 through 36.

Your inspection of results will of course be fundamentally different to the cards routine. Once your selection has been recorded you must now spin the wheel instead of turning over the next card. Though the time interval is therefore now much longer, very similar specific rules and scoring procedures otherwise apply.

But once you've again attained routine 75% scoring with roulette in private, you may then care to test out your skill in the more difficult casino milieus. However do not however expect a fortune here immediately. There are further problems in casino application when the moneyed chips go down. In the next Chapter I'll therefore deal with some of these…

A brief summary

Pre-call with roulette is all very similar to the earlier card skills. It may however prove more difficult initially - because this new skill is so definitely oriented into future time.

Here my investigations also revealed several new unexpected features, all readily predictable had I been more knowledgeable. These include lack of immediate transfer, between one routine and another rather different one. Interruption of training can greatly degrade your competence, as can a hostile atmosphere.

All such effects are however found frequently, in other more normal activities. As such they helped to confirm my growing impression of deliberate pre-call, as basically similar to most other high-concentration skills. Always of course excepting that prime feature, of time inversion that's involved. .

15

WHY IT'S HARDER TO WIN!

"Now listen to me Annette. Provided you are reasonable and have a head made of marble, cool and superhumanly cautious, there is not the slightest doubt that you can win anything you wish. And so I shall expend great effort to think and keep control"
Fyodor Dostoevsky - *The Gambler* - 1866

IN THIS CHAPTER I'll next deal with a proposition that may strike some of our readers as more immediate – the possibility of actually winning money in casino roulette. Here firstly my scoring at all relevant times can be taken as a dependable level of around 75% - assuming of course ideal conditions of full privacy.

The question was then how well this scoring level might hold up, under those more public or less studious conditions in the typical casino milieu. And the answer is that it can indeed hold up, though only after initial adjustment and always with greater difficulty!

Initially therefore I returned yet another dramatic drop in competence, when I first entered the estimable surroundings of the old Regency Casino, Glasgow. For a few evenings there I just stood apart from the table amidst the various spectators you'll always find in such places - recording my pre-call decision each time before the ball was thrown.

Then I would note the outcome about 30 seconds afterwards. This was of course yet a further large step into the future, as compared to the previous 15-second interval with my private wheel.

Diag. 15-1 therefore describes my early casino pre-call results – but crucially with no money placed. At first it was hard to adjust to the typical noise, bustle and raised voices of this new milieu. It was all so very much different, from those ideal and studious conditions of high concentration, which good pre-call competence requires.

But after a day or so my current high private performance began to reassert once more. In the second week I was back to strong scoring levels around 75% in this more public place.

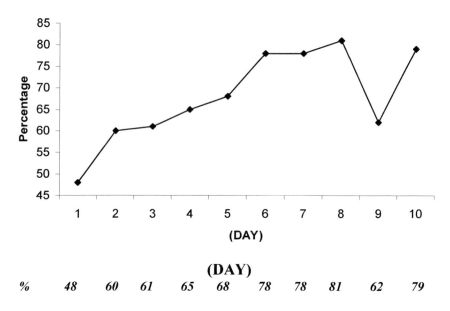

(DAY)									
%	48	60	61	65	68	78	78	81	62	79

Diag. 15-1: It took about 1 week to regain private roulette pre-call competence, in a more public casino milieu.

A quick and easy fortune now seemed very close! For if only I could hold competence to this level when the chips were down, I could surely double my money in great safety every day! Doubtless Dostoevsky, whom I've quoted above, would have understood!

For my next foray into casino-land, I therefore came armed with a bank of £100 to back whatever pre-call statements I would make. This was a fairly significant sum - approximately 1 week's average wages in Britain at that time. Very prudently also I decided to bet only in stakes of £2 apiece, meaning further that I had a very comfortable 50 reserves to start! Finally I decided to place all bets in double-binary mode – e.g. red-high, black-even, or similar.

But right from the start my prudent reserve bank – theoretically buffered to 10 times more reserves than required – began to shrink at an alarming rate! And the more my reserves became depleted, the faster my dwindling remainder began to shrink! I seemed somehow almost mesmerised into some downward spiral of failure: the more I tried to struggle against it, the worse things seemed to grow!

In the end I lasted only 6 hours spread over 4 consecutive days. I finished with just 173 winning decisions from a total of 392, a significantly bad - and indeed abysmal - 44.1% scoring average. This also meant a net loss of 46 units, made worse by a further 4 units lost to the bank when zero results ensued. In short this first attempt at casino betting had proved a total, unmitigated, disaster all round!

That my miserably below-average betting performance (44.1%), should contrast so totally with my non-betting competence (75%), seemed at first total mystery to me. For me it was still almost risibly easy to pre-call when no personal money was involved. But somehow the same task became much more difficult - indeed almost impossible - once the chips were down.

And of those many reversals of expectation I'd experienced with anti-memory, this proved quite the most mysterious (not to mention frustrating!), of all so far.

Inexperienced commentators would doubtless think, that the core challenge here lay in the actual possibility of pre-call, or reliable prediction of future outcomes. After which it should be just a tiny step further to the actual winning of money, through chips appropriately placed!

But in fact this problem is more the other way around. For proper experience can soon show, that pre-call (in any sensible situation) is a relatively easy feat. However the superficially simple addition of bringing money into the routine, can then degrade one's competence almost totally!

That there was something profoundly different, between these two situations, was therefore quite unavoidable. It was only years later, and after much perplexity, that I clarified a sensible analogy from more everyday affairs.

Imagination run riot

To pre-call at roulette - on its own and not contaminated by the extra stress of money – would therefore always be a mental challenge of some delicacy, one requiring an adequate degree of self-control. As such it rather resembles the somewhat similar challenge, of traversing a slightly slippery plank laid over a shallow stream. This is a feat you can likely achieve quite readily – providing of course you bring suitable care and concentration to the task.

In contrast imagine next the much harder challenge you would face with the exact same plank – but now laid over some raging torrent 100 feet below. *Physically* the actual traverse remains exactly as before. *Psychologically* however the challenge has now become much more formidable, indeed almost impossible for the untrained!

In the latter case imagination running riot is at work. But imagination' is just another word for our capacity to form mental images. And in times of stress such images tend towards the negative. On the high plank for example most people would start to imagine "How terrible to plunge down from here!" Which in turn provides a mental image or blueprint for failure, and so makes disaster more likely to occur!

So too it is with the great decrease in pre-call performance at roulette, whenever significant money is involved.

What others have said.

In this context English novelist Somerset Maugham (1874-1965) once offered an excellent, if rather impractical, betting system for roulette. You merely select the most worried looking player at your table – and bet the exact opposite to him or her all night! So that if he selects 'red' then you bet 'black', and so on indefinitely!

Although in practice Maugham's system may be rather problematic to execute, it still makes excellent sense from the anti-memory viewpoint. This is because worried punters should be all the more unfit to maintain psychological equilibrium under stress, so tending to "fall off the plank" and lose!

The great Russian novelist Fyodor Dostoevsky (1821-81), whom I've already quoted at the top of this chapter, was obsessed by something very similar. As a middle-aged writer worried over his finances, his gambling began in August 1863. He was then running away from his wife to join his young lover Paulina Suslova in Paris.

But first he stepped off the train at Wiesbaden to improve his finances at the local casino!

Soon Dostoevsky was 10,400 francs ahead, or about 75,000 dollars nowadays. This seems like a good example of that familiar phenomenon called 'beginner's luck'. He cashed in and left – only to have greed win out over reason as he was returning to the train. So the gambler in his personality dragged him back to the casino – predictably to lose most of what he had just won!

Over the next 8 years Dostoevsky then became a compulsive gambler around the roulette tables of Europe - always trying to capture and hold, that clear-headed attitude wherein he could usually win. His problem was that he tended to lose even faster, whenever he grew excited or confused!

From all of which came his semi-biographical novel *The Gambler*. Again too Dostoevsky's tale makes excellent sense in pre-call terms. It's a story of dimly realised paranormal ability, but often degraded into psi-missing by stress overloads….

More refined ventures

From time to time as occasion offered thereafter, I therefore made several progressive attempts, to resolve my failed challenge of roulette in moneyed mode. Bit by bit I introduced slight changes and new measures intended to strengthen that ever-labile pre-call capability. For example I soon limited my efforts to just 1 hour daily, a maximum schedule of 50 double-binary trials or 100 individual bets.

Further I introduced a daily stop-loss of 20 units – the most I was prepared to lose on any particularly bad day. My starting reserves were also strengthened to 200 units, conferring a strong endurance of 10 such days.

In this manner during the mid-nineties I gradually built up to £10 units and an average controlled competence around 53.5%. Allowing for zero, this meant an average profit of just over £50 or 5 units every notional day. This was obviously lamentably poor performance compared with my non-betting 75% competence. Still it was better than those gross margins on which casinos operate so profitably

But now a further and final problem intervened. Each time I tried to raise stakes beyond their customary £10 level, my current and curiously consistent competence of 53.5% began to abrade away again. Or, in terms of my previous analogy, the plank was being raised too high once more!

To reinstate and maintain control at this higher level would require near-professional dedication, and more perseverance than I was now prepared to give. For I was never really interested in practical roulette anyway, rather the pre-call aspects of the same.

To others with more time, dedication, self-control, I therefore relinquished the challenge of winning consistent money in casinos, and conquering roulette more competently than I have ever done. That the feat can indeed be achieved I have no doubt at all.

And for the benefit of any such future aspirants, I'll now highlight several further relevant factors, in the typical casino milieu.

For a start one must realise that the physical conditions are hardly ideally suited to any high concentration exercise. There's too much pushing, shoving, extraneous noise from croupiers at different tables, interjections from punters beside you, and so on.

In addition the *pace* of the typical British or American roulette session is always highly variable. Dependent on croupier whims and how busy a table is, this pace can vary between 0.5 and ca. 3 minutes, for each individual spin.

In addition there can be strong interpersonal transactions – mostly subliminal and unrealised - between the player and the croupier. These constitute another significant psychological variable, in terms of Child-Adult transactions, as I've already explained (p.154).

All of which further means that some croupiers, are definitely harder to win from than others, at most times. This is an observable fact, which sceptics without practical experience would probably decry. And yet most punters - and further all casino managers - well know this croupier effect. But without ever really understanding it properly.

That's why it's standard casino practice for the management to change a croupier, who exhibits too much empathy (i.e. parallel interpersonal transactions) with a punter winning steadily. Which again of course makes perfect sense in terms of that interpersonal diagram I've depicted earlier.

A whole book then could - and maybe should - be devoted to this greatly tangled, though usually over-simplified, problem of casino gambling. One part of it might clarify how a great many of its supposed 'superstitions', make excellent sense in pre-call terms. Another might well analyse the pre-call potential in more detail….

Automation is easier!

In recent years also there have been two new roulette develop-ments which can make it easier for pre-call to operate. Firstly most casinos, in Europe at any rate, have introduced new automatic tables. These greatly diminish those old annoying variables of pace, crowding, pushing, etc.

Instead such tables combine mechanics and electronics for the benefit of some dozen players seated all around. The mechanical aspect involves the ball – now shot out by an automatic spring mechanism – still circling round the traditional counter-rotating wheel. The result is then read and recorded by electronics – which also keeps track of each punter's cash and record on a private monitoring screen.

These new electro-mechanical machines have therefore completely transformed the old croupier and table situation, while remaining every bit as open and transparent as before. From the pre-call viewpoint their great advantage is to have minimised, those several interpersonal variables of psychological complexity.

Instead they allow you to sit in large privacy, at your own screen console. The ball always starts up, every 60 seconds or so, at a regular rate of spin. Both these routines are psychologically favourable to any high concentration skill.

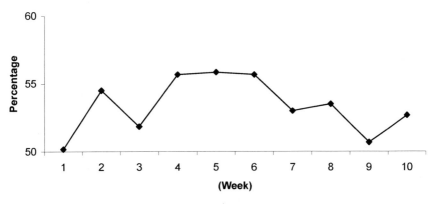

SCORE/600	301	327	311	334	335	334	318	321	304	316		
PROFIT(g)	+2	+54	+22	+68	+70	+68	+36	+42	+8	+32	=	+402
ZERO (-)		-5	-11	-10	-6	-9	-5	-11	-13	-7	-8	=	-85
PROFIT(n)	-3	+43	+12	+62	+61	+63	+25	+29	+1	+24	=	+317

TABLE 15.2: Results over 10 weeks with electro-mechanical machine.

A few years ago therefore, while researching this book at the British Library, I decided to try out these new automatic tables at several casinos around Central London. My starting bank was precisely £1,000 - i.e. 100 x £10 units, which I bet in double-binary format (e.g. red-high) as before.

I placed 60 such bets (i.e. 120 individual trials) more or less consecutively, an effort that took just over 1 hour per day. And I followed this regular schedule 5 days a week for a total of 10 weeks, so making 6,000 individual bets in all. This total excluded zero results, which were as usual handled separately.

My final score was then a highly significant 3,201/6,000, or an overall average of 53.35%. This meant a *gross* profit of 402 (i.e. 201x 2) units overall – which at £10 per unit meant £4020!

But from this gross return I had also lost 85 doubled zeroes, which the bank claimed as its advantage – i.e. 85 units or £850. So I ended up winning a *nett* £3,170 overall.

There were however several downsides to this promising little experiment. For one thing my proven competence was only about one-eighth as good as it should have been ideally. That 53.3% scoring - which I attained in practice - meant only a gross advantage of 6.6%. But 75% scoring - which in theory I should have been able to attain - would have meant a far more robust 50% advantage to my side.

In addition my performance peaked, again at a curiously stable level of 55.7%, between weeks 4 through 6. After that it began to degrade down again, presumably because I was now tiring of the strain. Still I ended up with that curiously consistent scoring level of 53.35 % overall – again strangely close to that similar total (53.5%) I'd amassed several years before.

It therefore appeareed to me as if some subconscious mental block or limit was holding me down at this level at most times. It all seemed a most curious consistency, on which sports psychologists might perhaps have something useful to say.

My betting performance was also always relatively dismal, as compared with my true pre-call capabilities. Very likely any competent golfer or darts professional - or indeed many others accustomed to high concentration sports - could have easily surpassed me, in this particular application of pre-call skill....

That roulette in at least two forms, can nevertheless be conquered to win money, is in any case quite clear. So that Einstein was wrong, and Dostoevsky right, in this context.

Finally this new concept - of roulette as a game of skill instead of pure chance – may well be derided, by mathematicians or statist-icians who lack practical experience. Still it would probably be more acceptable, to those with actual experience of the game!

Enter the Internet

The second new form of roulette now available, has also its good and bad points as before. On the ubiquitous Internet there are now some 2,000 virtual 'casinos' – all offering roulette to punters willing to place money via their debit/credit cards. So that you can now sit alone, in the *total* privacy of your own study, with a virtual roulette table all ready for action on your own computer screen. The pace of the game is also entirely dependent on your own inclination, whenever you next decide to bet on choice!

At first glance therefore these Internet casinos offer psychological conditions even more ideally suited to a high concentration exercise. But again there are some serious defects.

For example the ball you see circling your on-screen wheel is really just virtual reality or electronic illusion, entirely governed by a computer with an electronic Random Number Generator (RNG) device.

This further means that each individual outcome is decided by electronic state transitions deep within tiny computer chips – about as far removed from the everyday experience of intuitions as it's possible to be. Indeed it's at equally far remove, from that likewise human observation of the spinning ball and wheel.

All in all therefore, to assume that pre-call should also apply in this purely electronic situation, must involve methodological extensions of some magnitude. That new and unknown variables may then be involved, at one or several stages in this illusory electronic roundabout, is at least quite probable. The whole thing requires much more testing at this stage.

In addition there's always the danger that your anonymous site operator could easily switch over, to a self-defeating glitch program, if ever you started winning seriously. Which can hardly inspire total confidence in what you're trying to do, but rather must conspire against that ever labile pre-call skill.

A first safe rule for the wary punter must therefore be to choose a reputable Internet casino site. The best are those run by the major British bookmakers, to whom reputation is all. These all offer European roulette with just one zero, which means a casino deduction of -2.7%, for even chances overall.

But true roulette afficionados can always go further, by searching out French roulette where the casino advantage has been halved again, reducing (for even chances) to just 1.35% on average. However French roulette is usually offered in low profile, so that it can take some digging to find a reputable game.

Application of pre-call to Internet roulette therefore remains at present unproven, and never at all investigated suitably by me. It would be a task requiring considerable focussed dedication overall. But for those who would still like to try it out, those learning instructions I've given in Ch. 14 always afford that open invitation to "Try it and See!"

Meantime in my own case I had more important things to do....

A brief summary

Though casinos are hardly conducive to delicate concentration, pre-call of future outcomes can be made to function there too. This also suggests that many practices commonly dismissed as superstitions, in fact make excellent sense in terms of repressed anti-memories.

But, as with other high-skill exercises, the pre-call task becomes much more demanding as serious money gets involved.

Electro-mechanical machines are less variable of procedure, so that pre-call in betting mode is easier with them. Still it's an exercise which must demand almost professional dedication, and again liable to get harder as the stakes get high.

Good Internet casinos may be more favourable again. They appear to offer even more ideal conditions of total privacy. They can function at any time, place, pace you care to choose. More research is however required to see if winning in this mode is feasible.

16

ELECTRONS ARE PREDICTABLE

"The position you hold on the quantum reality issue, is more like a religious conviction than a matter of science."
John L. Casti – *Paradigms Lost* – 1989

The Copenhagen Doctrine
A more important new mode, for testing the limits of anti-memory capability, was obviously quantum theory. This is the set of mathematical rules that describes how electrons and similar particles behave. As a mathematical formulation, quantum theory has long proved to be a total physics success. That X-rays, lasers, TV, computers all now work is practical proof of its efficacy.

More philosophically however quantum theory has one great defect. Nobody knows exactly what its rules mean *in reality.* Or in other words its ontology – what it can tell us about the basic structures of nature - seems woefully confused, self-contradictory, unsatisfactory, incomplete.

This problem began to surface immediately after two pioneers had produced two different mathematical routes to the final quantum theory in 1926. Erwin Schrodinger (1887-1961), and Werner Heisenberg (1901-76) were immediately taken up by Niels Bohr (1887-1951), who was always the leading quantum theorist. Through their descriptions of electrons, he successfully clarified the Periodic Table of all elements.

Still Schrodinger's equations can only afford the *probability,* of an electron being in any particular place. So they can only describe an electron as being maybe *Here,* but possibly *There* - and likewise for all points between! This description was also reinforced by Heisenberg's *Principle of Uncertainty*: you can never be fully certain, of *both* the position and velocity of an electron, at any particular point in time.

Instead of it being a definite particle, quantum theory therefore treats the electron as more like some nebulous cloud of mist, one smeared out with different densities all over that region where its equations rule.

But this possibly smeared-out nature of the electron, can only hold true, up to a certain definite instant in time. For experiment shows that the electron exists as a definite particle whenever it's actually observed. Supposedly then – through a process dubbed *quantum collapse* - its nebulous nature has (somehow!) collapsed back again, to make it a definite particle at that particular time and place!

To explain all this Bohr proposed his principle of *complementarity* – the idea that an electron can be considered in two different ways. From one viewpoint it may be considered as a particle, from another as a wave. The important thing is that these two viewpoints can never be applied *simultaneously.*

Bohr also insisted that this strange alternation, as partly described by the wave equations and differently by observation afterwards, does indeed describe the electron's real nature. If so, it must sometimes exist as nebulous, probabilistic, or smeared-out like a fog in actuality. But immediately some person (or even a machine) actually observes it, the electron supposedly collapses, back into a definite particle!

This was the essence of the Copenhagen Interpretation of quantum reality, proposed around 1930. It took its name from the Danish capital, where Bohr had his headquarters in an elegant mansion donated by the Carlsberg brewery firm. And such was his reputation that it soon became widely accepted, by the general science community.

Thence it found its way into student textbooks, where it remains down to the present day. Likewise it has spawned a strong nihilism in popular culture, one which regards all nature and life as basically unpredictable.

But physicists are basically pragmatists instead of philosophers. So mostly they've just used the quantum mathematics with great success, while seldom worrying too much, about what it all may actually mean.

In more ordinary terms however, the quantum mathematicians may usefully be compared, to blind croupiers at a roulette wheel. With great exactitude they can calculate the probability of red-or-black *totals,* which may turn up next over any number of trials. But still they can never predict any *singular* result - either through their mathematics or through new mental capabilities they've never realised.

And yet these blind croupiers insist that their statistics reflect the true limits and reality of nature, and not just the human limitations of their own routines. Which might perhaps usefully be considered in terms of ' Inverse Expertise', as I've proposed earlier in Ch.14.

A contra-intuitive idea

But whether or not these indefinite quantum equations reflect the true situation in reality, has never at all been clear. For mot only do they conflict with the strict determinacy of all previous physics, they're also some of the most contra-intuitive ever seen.

So that right from the start not all the great physicists agreed with Bohr. One notable opponent was Einstein, who held that the quantum mathematics were merely a half-way step to the truth.[1] Prince Louis de Broglie, who had earlier uncovered the electron's wave properties, was at first likewise doubtful:

> "The construction of purely probabilistic formulae that all use today was thus completely justified.
> "However the majority, often under the influence of positivist doctrine, have thought that they could go further, and assert that the uncertain character of knowledge at its present stage, is the result of a real indeterminacy."

Later however de Broglie reversed this stance under the influence of W.Pauli. The latter of course tried to relate quantum indeterminacy to what he saw as the equally indeterminate nature of paranormal experiences, as we've already seen under *Synchronicity* (Ch.8).

There were also various paradoxes that questioned Copenhagen validity. Of these the curious problem of Schrodinger's Cat became best known. He imagined his fabled cat in a poison trap which can be set off by some singular quantum event - like the entrance of just one electron or beta-particle. Supposedly then one must regard the cat as *both* alive and not-alive until its fate is inspected, that is we grant reality to the quantum rules![2]

Later however it will be *either* definitely alive or definitely dead - when the trap is opened by some observer to see which is really true!

Absurdity rejected

Famed British astronomer Fred Hoyle also extended this Schrodinger paradox into absurdity. He replaced the mythical cat trap, by a quantum detonator for an atom bomb, placed in some city like London. Supposedly then London must be regarded as *both* destroyed and not destroyed by the rest of us outside – that is until some bold observer ventures to inspect the actuality![3]

Hoyle therefore dismissed the entire Copenhagen Interpretation as demonstrable nonsense, a position with which most commentators now probably agree.

Over the last 20 years therefore, there has then been a growing reaction against Copenhagen mysticism, and its implication that the tiniest workings of the natural world, are ultimately governed by sheer chance. This is an inference quite at variance with all other physics laws before and since.

While the mathematical formulation of quantum theory is therefore still totally reliable, Bohr's Copenhagen Interpretation of it seems largely indefensible. Many now hold, with good reason, that it was demonstrably unscientific, in its extension from theory to reality.[4]

These new sceptical realists think it absurd that our understanding of nature, should depend on such elusive imprecisions as super-position, wave collapse, uncertainty. So that few really try to defend the Copenhagen idea anymore.[5]

Yet how to construct a new ontology from those marvellously successful quantum theory rules - a better idea of what they can all mean in reality – remains problematical.

In his book *Paradigms Lost* (1989), physicist John Casti then lists over half-a-dozen alternatives. Just which of these alternative ontologies, or quantum pictures of reality, you may then choose to adopt is currently more a matter of faith than science, as Casti says. For there's no pragmatic test available to decide between them all.

Handshakes across time!

Two of these other rival quantum ontologies are also quite determinate. One is the *pilot-wave* concept proposed over twenty years ago by David Bohm (1917-92) from the University of London. Alternatively there's the *transactional* theory proposed by John Cramer at Washington University.[6]

Bohm treated the electron like a classic particle, which always exists and oscillates in waves. (No nebulous mist of non-locality there!) These waves determine how the electron should behave. The front or pilot part of the wave sends back messages of what lies ahead. These keep the electron as a classical particle, which is always quite determinate, with no uncertainty involved.

Unfortunately however this retro-transmission of information requires faster-then-light signals. And these current physics most definitely forbids. But otherwise Bohm's theory does far less violence, to our impressions of reality than Bohr's Copenhagen idea.

An equally determinate alternative is John Cramer's Transactional Interpretation advanced ca.1985. And this is even more intimately concerned with time. It questions a curious imbalance accepted into physics over a century before.

This imbalance took root in 1873, when James Clerk Maxwell formulated his famed basic equations for electromagnetism and light. From these all kinds of current radio broadcasts – wireless, TV, radar, mobile phones – later evolved. But Maxwell's equations stay unchanged when the time direction is inverted or reversed.

They have then two equal and symmetrical solutions – a retarded potential and an advanced one. In radio terms the retarded potential is the one we hear customarily - a broadcast from the past up to Now.

These retarded waves also exhibit *causality* - the common idea that every event which we label *present,* has been caused by another which we label *past.*

Conversely the advanced potential would mean a broadcast from the *future* back to the *present*, and so fundamentally reversed in time. As such it would have to involve *retro-causality,* a possibility now considered seriously by some physicists.[7] With retro-causality an event which we label present, would have its cause in a future one.

So far however advanced waves never seem to have been observed in reality, so that pragmatically they've been ignored.

In addition John Wheeler and Richard Feynman – two of the 20th century's greatest physicists – proved mathematically that 'advanced' broadcasts (from the *future*), should always be totally negated by the more usual 'retarded' ones (from the *past*). They should be cancelled or neutralised because they're exactly out-of-phase. However this proof dealt with a very restricted situation, not at all as complex as reality.

In any case Cramer has adapted this thinking into subatomic quantum realms. He applies *both* advanced and retarded solutions to electron quantum waves. Each quantum event then becomes a bit like a handshake – a transaction across space-time in keeping with other physics conservation rules.

If so, at this very tiny or subatomic level, time seems no longer like a one-way street, one wherein traffic always flows from past to future realms. Instead it's more like a two-way street where the past determines the future - but also vice-versa!

This treatment is fully consistent with Einstein's Relativity Theory, and resultant Block Universe ideas that I'll describe in Ch.18. These see reality as totally predetermined at all times. But still Cramer's two-way transactions permit human freedom of choice – providing of course that the laws of physics are obeyed.

This last is also important in the more general context of free will. For example you're always free to choose to jump 100 feet straight up from the chair you're sitting in. Except that Newton's law of gravity (and also the presence of the ceiling!) would then forbid you from doing so!

That there are various less surreal alternatives to the Copenhagen Interpretation is then quite clear. Still nobody so far has ever been able to suggest any experiment, which might discriminate between them all.

Partly this may be because of that dichotomy of specialisation, or *Inverse Expertise* as I've termed it earlier. So that highly specialised physicists, can hardly expect to be equally knowledgeable in matters of the Mind. Wherefore they've tended to assume that the range of mental capabilities with which they calculate, is really all relevant that there is to know.

But in this they may resemble their forbears up to about 1870, who assumed that our range of colour vision, must span the totality of the light spectrum.

Pragmatically in any case Bohr's doctrine does enshrine one very definite consequence. It entails quite clearly, that the next action of any one singular electron, must always be unpredictable *in principle.*

In this it differs fundamentally from other future-oriented events like a roulette spin. The latter result is always predictable in principle. So that you could calculate exactly where the ball will land, if you were to measure precisely its position, velocity and the wheel's revolution rate. As indeed some punters are now applying computerisation to achieve.[8]

At this point too the time-extended capabilities of anti-memory grow relevant. So could the singular behaviour of an electron ever be pre-called like a roulette spin? Even though Bohr's interpretation, based on a more limited span of mental capabilities, would claim that this must be impossible

Here then the Copenhagen Interpretation might perhaps prove to be *falsifiable*. Falsification is a concept we owe to philosopher Karl Popper (1902-94). He clarified that any scientific theory is only as good as the next experiment, which might perhaps prove it false.

Is an electron predictable?

The crucial question is therefore whether the singular behaviour of any electron, is really unpredictable after all. The Copenhagen Interpretation states very firmly that this must be so. So that *if* the electron's future were to prove as predictable as a roulette ball, this particular interpretation would have been falsified to that degree. Whereas alternatives like Cramer's Transactions would have been strengthened conversely.

However this Copenhagen consequence has apparently only been tested out by direct experiment just once. That was in 1969 when ex-Boeing engineer Helmut Schmidt, worked briefly with J.B.Rhine who promulgated ESP (Ch. 7). He constructed an RNG (Random Number Generator) based on a Strontium-90 radioactive source. Periodically it lit up 1 of 4 lamps in a wholly random sequence, one that the Copenhagen Doctrine would claim as *in principle* unpredictable. [9]

Nevertheless Schmidt did find two people whose intuition enabled them to outguess his quantum generator, consistently if marginally. For example, in a total of 22,569 trials, one of them returned an overall score of 26.3%, where 25% was the expected chance result. The statistical probability of this result was about 10^{-6}, or the chance odds about 1 million to 1 in more common terms.

A similar manifestation of electron unpredictability, is that you should never be able to predict the *exact total* of electrons, which will shoot out in the next 10 seconds, from that radioactive Strontium-90 source. Would pre-call therefore again function in this situation, in partial replication of the Schmidt experiment?

Pre-call again possible

To answer this question I provided myself with a weak radioactive source of Strontium-90 – one shooting out thousands of electrons (or beta-particles) every second, in a great haphazard stream. I then set up a Geiger counter in front of it, so counting up all those electrons emitted over a typical 10-second interval.

Experiment soon showed that some total like 32,174 emissions per 10 seconds would be more or less standard for this set-up. With such large totals also the last two digits can be taken as entirely random – and quite unpredictable according to the Copenhagen idea.

I further decided to classify these last 2 digits in double-binary format - just as with cards and the roulette wheel before. So that for example, if the final two digits were 12, I would regard them as 'low-even'; a presentation of 83 I would classify as 'high-odd' likewise.

The question then was whether pre-call would again function in this fourth learning mode, and much as with the other three variants I'd tried out previously?

It did not take me take long to conclude that very probably it would. I therefore decided on a formal series of 5,000 double-binary trials - to be conducted in a series of 50 sessions over the next 10 weeks.

As with my three earlier pre-call routines, some re-learning was again required for this fourth new learning mode. But this time it was all much easier than before. A opposed to total learning of some new skill, it felt more like regaining accustomed competence in a slightly different mode One does learn to pick up new skills faster, the more experience is gained.

By the end of this series in any case, it was then quite obvious that radioactive quantum totals could be pre-called. Further it was all very much as with cards and roulette before. In its secondary features also (e.g. long winning streaks at first too quickly countermanded by an opposite), this fourth mode of pre-call seemed little different from the previous ones.

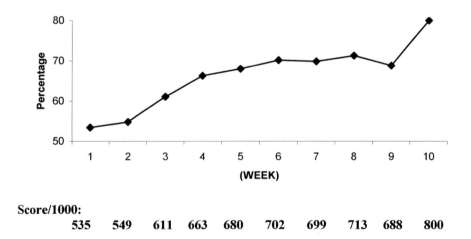

Score/1000:
535 549 611 663 680 702 699 713 688 800

DIAG. 16.1: Pre-call of beta-particles was very similar to roulette scores, in both competence levels attained and secondary effects.

All of which also of course totally contradicted the conventional Copenhagen Doctrine, while confirming those pioneering findings of Helmut Schmidt 30 years before!

As always also, my findings here are not presented as any sort of definitive experiment. Instead they're more of a model for emulation, one which others can always test out for themselves if so inclined.

In fact there might well be a Nobel Prize here awaiting any suitably visionary young physicist, who cares to attempt replication of this experiment more formally. His or her requirements for doing so, would merely be a sensible measure of everyday psychology, the kind I've earlier termed Shakespearean. Plus a minimum previous 100 hours of direct pre-call experience, in one of those simpler earlier routines...

To say that I was shocked and astonished by these latest quantum findings would be to put the matter very mildly indeed. After all with cards there had always been suggestions in the paranormal literature, that they might be 'guessed' (correctly!) if properly approached.

With roulette too there were plenty of so-called 'superstitions', which mostly made sense from the more enlightened anti-memory viewpoint. There were also plenty of anecdotes about anomalous luck in gambling – like the strange sequence of "the man who broke the bank of Monte Carlo" in reality a century ago.[10] And in any case there was always conventional physics to state that the fall of the ball should be predictable, in principle at least.

With electrons however there was the great negative convention of Copenhagen, proclaiming through thousands of textbooks that singular prediction just could never be! Whereas pre-call was now demonstrating, that electrons were every bit as predictable, and so probably determinate, as larger particles.

To the clear Childlike gaze of pre-call, it now seemed obvious that this particular Emperor of Copenhagen, had in reality no clothes at all!

Later in any case, as my initial shock receded and I adjusted to the new facts now readily observable, the central issue here slowly clarified. For while Bohr and his cohorts were undoubtedly pre-eminent in physics, they never seem to have scrutinised their own undeveloped mind capabilities at all. Though perhaps W.Pauli, who considered the paranormal possibilities along with C.Jung (Ch.8), should be regarded as the sole exception here?

But had any of those early quantum theorists been expert in pre-call competence, they could easily have subjected their curious quantum contentions to the acid test of direct experiment. And so pragmatically discovered that they lacked validity!

In which case also micro-time (in the quantum world) would hardly have seemed so different, from middle-time (in the everyday world of direct experience). Both regions would likewise be governed by the same kind of determinacy found in non-quantum physics laws. So that current physics might perhaps grow more unified in this way.

For, as the journal *Science* has summed up our current under-standing in all these affairs:[11]

> "In short, quantum mechanics, special relativity, and realism cannot all be true"

A brief summary

Quantum theory is a triumph of modern physics, with the entire world of electronics governed by its rules. Yet, amazingly, what it all means in reality is still very much obscure. To explain this the Copenhagen Doctrine was advanced some 75 years ago. It held that the singular behaviour of an electron must always be unpredictable.

However few theorists believe in this doctrine nowadays, and several more determinate alternatives have been proposed.

The weak point in Copenhagen ontology has always been too limited concepts of Mind capability. Its theorists never considered that antimemory might occur. But now through a simple - and apparently readily repeatable - exercise of pre-call, it seems that their doctrine is quite falsifiable!

The required inference must be that micro-time in the quantum world, isn't really all that different from middle-time in the world of everyday experience.

So the important next question is whether micro- and middle-time, conjoined into similarity as anti-memory now shows them to be, might not further be unified with macro-time. This last is 'deep time' – those billions of years apparent throughout the history of the Universe. Macro-time is described by Einstein's two Theories of Relativity. And, as we will now see, it also provides a localised setting, for a fifth pre-call routine....

PART FOUR:

ANTI-MEMORY: WHY IT CAN BE

*"(Science needs) a meaningful isomorphism,
between the relationships of time
in experimental psychology,
and the description of time in physics".*

J.T.Fraser – *The Voices of Time - 1968*

17

DOES TIME REALLY 'PASS'?

> *"One suspects then, that time doesn't really 'flow' at all; it's all in the mind... I maintain that the secret of mind will only be solved when we solve the secret of time"*
> Paul Davies – *God and the New Physics* – 1985

THAT INTUITION IS indeed a real and valid function of intellect, though one largely misunderstood, had now for long been clear to me. That it's also a highly censored faculty was also observable, indeed predictable. For internal censorship is only to be predicted, when unexpected novelty conflicts too much, with what we like to think, we know about the world.

That intuition can then be developed, in any number of modes numerate and otherwise, is a further finding I've described. The method is always deliberate practice, aided by some familiarity with the intuition process, and due understanding of its subconscious origins. This matter I now regarded as proven to myself, and further provable by anyone so interested.

Finally that intuition is best *described*, in terms of pre-call or anti-memory, seemed also very clear. All those old former paranormal confusions – *'telepathy'*, *'clairvoyance'*, *'remote-viewing'* *'ESP'*, etc. – can then be traced to poor observation of the natural facts. Along with that common priority, which makes most people think in terms of space, before considering time.

Nevertheless, when all was said and done, my four new pre-call modes – binary, three-digit, roulette, electrons as described in Part Three - still seemed to me quite small and puny temporal experiments, at most extending only 1 minute into future time.

Which was also obviously a very brief and tiny interval, as against those much longer time-spans apparent in everyday intuitions, or indeed against which we live.

What I therefore required next was some larger, but still numerate, situation from real life, against which pre-call might be tested out and gauged. This was my required *fifth* mode of pre-call, which I eventually did discover, only to find that it was by far the most wondrous and indeed astonishing of all. But that is best left until Chapter 19: more meantime is to establish how pre-call or anti-memory could ever possibly occur.

And that is a question intimately bound up with one of life's great questions – the 'real nature' of time. This problem of time has also always been subject to two very different interpretations, as we'll see.[1] After which it may seem that, far from contradicting the nature of time in reality, anti-memory can really support the latest understanding very forcefully!

Time's cultural origins

In our western culture from which science has evolved, people have always regarded the future as open to choice and free will. To them the *present* seems forever changing, always (somehow!) transforming the formless *future* into the more structured *past*. Which seems to imply that time is always (somehow!) flowing past us, like some great eternal and metaphysical river ceaseless in its course.

Such thinking also strongly depends on the Indo-European school of languages, through which most western culture and science has evolved. Indo-European languages are uniquely *tensive*. This means that, compared to other languages, they're unusually replete with verbs we can tense along with temporal adverbs- so describing *past, present, future* in many refined modes.

Likewise our other temporal or tensed words – like *will, soon, did, always, ago* - all force us unwittingly to think about time in certain ways. Though they're mostly taken on trust from the thinking, of those shadowy Indo-Europeans 8,000 years ago!

The same was true at the start of western philosophy, which began at Miletus on the Aegean around 600 BC. There the new Greek rationalists, who loved to debate the nature of all things, were soon pondering the puzzling properties of time. One such was the early philosopher Anaximander (d.550 BC), who thought of events as 'coming to be' and then 'passing away'.

Anaximander preceded the more famous Heraclitus from a century afterwards. Heraclitus saw *change* as the most essential feature of this world. Hence came that potent allegory of time as some great eternal "river of no return" - a theme which has endured in romantic ballads down to the present day.[2]

Through this allegory of the river, time seems forever flowing past us, bearing away all events from its inaccessible upstream *(future)* into its equally unreachable downstream *(past)*. But does this mean that events are already pre-formed *before* we encounter them, just as they are more surely rendered forever unchangeable, *after* they've passed us by?

"You can't step in the same river twice!" said Heraclitus. By which he meant that you can never go back to relive, redo or even revisit your past life. "Nor even once!" retorted one of his followers. For even the single act of immersion must inevitably make you somewhat different from before!

Eventually such views became formalized as the philosophy of *Becoming,* the doctrine of ceaseless change and transience. Broadly its adherents see *present* reality as in a state of eternal flux or change. The *future* is either nothing at all, or else an unpredictable time region – one quite open to manipulation by present choices we can make through free will.

Though some would hold that we can't really change 'the future' – since it seems best defined as merely 'what will be'!

In any case, and at some mysterious regular rate, this personal *future* seems also forever transformed, via the changing *present*, into an opposite region we term the changeless *past.* The *past* is that time sector which has somehow *passed* by us, or the sum of events embedded therein,

In size the past of the Universe is also very extensive – about 15 billion years in duration if we are to believe current cosmologists. Happily too the future seems even more extensive – with at least 100 billion years left to the cosmos according to the same authorities. So that in reality the Universe is just a teenager in human terms!

All of which also highlights the *present* as a curiously brief interval in between. Perhaps it has no duration whatsoever, being merely an interface between those extensive past and future realms. *"What we mean by 'right now' is a most curious thing"* - as physicist Richard Feynman once observed.

The Ancient Greek philosophy of *Becoming* also found support in the Jewish Old Testament. For the latter afforded a firm record of linear change or progression, from the first Creation moment in *Genesis* down to much later times.

And where the Jewish religion had thus put a firm start to time and all later events, the Christians rather tidily concluded it as well! They believed that all time would come to an end for humanity, on that greatly dreaded Last Judgment Day!

Christian philosophers were however much troubled by two attributes they ascribed to God's perfection – omniscience and omnipotence. For if God was so all-wise as to know both past and future, how could He then also be all-powerful and change future happenings at his will?

In any case from such sources Western culture derived its notion of time as a linear progression from *past* to *future*, with the *present* as some transient and transforming interface between. This supports our' comforting sense of egocentricity: it's we who endure while time (somehow!) "passes by".

And it all relates - if a trifle vaguely – to Heraclitus and that great allegorical river of no return...

Newton's mistake

Reflecting its common acceptance into western culture much earlier, this everyday notion of time as somehow 'flowing', was also formalized into the newer science culture by Sir Isaac Newton. He did this in *Principia,* (1687) his great book of physics wherein he famously defined time as absolute: [3]

> "Absolute, true, and mathematical time, in and of itself and its own nature without reference to anything external, flows uniformly and by another name is called duration.
> "Relative, apparent and common time is any sensible and external measure of duration...and is commonly used instead of true time."

By Absolute Time Newton meant the entire temporal summation, of all events throughout the Universe. This also implied an extension of common thinking about the *present*: the moment of *now* became a single entity which must extend into all places everywhere. So that if your watch is now showing 12 o'clock mid-day wherever you happen to be, it must also be noon in every last corner of the furthest galaxy.

With Newtonian time each passing second of the *present*, might then be compared to some metaphysical new line of bricks, added on top of the equally metaphysical wall of the *past,* a wall extending throughout the entire Universe. Continually also new bricks are being picked up from the unformed piles of the *future,* and then cemented by ourselves or Nature on to that more rigid structure of the *past*.

Apart from ourselves, and our presumed free will, this process is also quite determinate. So that if you were wise enough to know all the operating laws of the Universe, then you could also predict its entire future, which must be predetermined at all times.

Newton's idea of Absolute Time, was also linked intimately to his other idea of Absolute Space - a theoretical space framework for the entire Universe. It would be a sort of scaffolding which you could always measure – by its 3 traditional dimensions of length, breadth, height - from some fixed corner point somewhere.

Philosophers however soon realized that Newton's concept of Absolute Time must be mistaken fundamentally. For if time is regarded as somehow flowing, this can't be *"without reference to anything external"*, as he said. There must always be something else with reference to which it flows. Or if one regards time as some allegorical river forever flowing past us, then where or what are the banks by which it is contained?

Nevertheless Newton's ideas worked well enough in physics for the next 200 years. And so they have continued to reinforce those common western notions, about time as (somehow!) always flowing past ourselves. The past is closed but the future is open to choice and free will, so leaving the present as that transforming interface between. All these are poorly based notions which continue to inform most people, and which would of course forbid any process like anti-memory.

Likewise when formal psychology began to shape up as a separate science after 1850, it too incorporated the same general idea of Absolute Time. The same was true of parapsychology when it started under J.B. Rhine around 1930.

Both disciplines continue to see themselves as ideally a branch of physics, though still clinging obstinately to those old common or Newtonian time ideas. As such, once again they could hardly accommodate the possibility of anti-memory.

People then often state as a matter of faith that we can never know about random events in what they term the future realm. Still they can hardly state precisely why this should be so.

However, the truth is really not at all that clear. For all such thinking depends on that *Prime Assumption* I've discussed earlier (Ch.4). People have always just simply *assumed* that memory must inherently be limited, to direct contemplation of the past alone. They've always accepted this belief as self-evident, but without ever making the slightest check-up, on whether it was really true.

And yet that slight 10^{-7} anomaly, of intuitions or anti-memory in everyday experience, suggests that in reality things may be otherwise. Given such, all theories and beliefs derived from *The Prime Assumption* are equally suspect…

A brief summary

The common, though vague, idea that "time passes (by us)" stems from Ancient Greek philosophy, and was thence adopted into the western cultural ethos. Thereafter it was formalised into science by Isaac Newton three centuries ago. As such it's still totally accepted throughout psychology today.

Such thinking however all rests on one *Prime Assumption* not previously realised – i.e. that Mind or memory can just con-template the past alone. But this previously untested assumption is one which applied anti-memory can now readily falsify, and so challenge fundamentally.

18

OR DO WE 'PASS THROUGH' TIME?

"There can no longer be any objective and essential division...between 'events that have already occurred' and 'events which have not yet occurred'."
C. De Beauregaard - *The Voices of Time* - 1968

The philosophy of *Being*

IN ANCIENT GREECE there also existed a very different time philosophy, one diametrically opposed to *Becoming* which Heraclitus espoused. Instead of his view that time is (somehow!) forever "flowing past" us, it held that we are (somehow!) just "passing through" reality.

This was the time philosophy of *Being*. Its most famous exponents were Parmenides, and his pupil Zeno, who lived at Elea in Sicily around 400 BC. Adherents of *Being* maintained, that since reality was essentially static, if anything moves at all it can only be our changing states of Mind. To prove this point Zeno advanced 4 arguments, ingenious paradoxes which philosophers ponder still.

To a certain extent also this ancient debate reflected the more obvious question of the passing sun. Since everyone could see that the sun goes round the sky once daily, most people accepted that the Earth is static while the sun passes by. Only a few would have dared to consider the converse – that it's really our own massive Earth that passes (at about 60,000 miles an hour!) around a relatively static sun.

Likewise most people accepted the doctrine of Heraclitus - that time is (somehow!) always 'passing by' us all. Few could have dared like Zeno to contemplate the opposite - that it's really we who are (somehow!) moving through static reality.

And there matters largely rested for the next two thousand years. Reinforced as it was by common impressions and the major Judaeo/Christian religions, the notion of time (somehow!) 'passing' by us, remained a dominant theme in western thought.

The Block Universe

Over the last two centuries however, *Being* began to revive again. Partly this was because the rapid development of technology was offering new models for time thought. The common feature of such new models is that we humans are just "passing through" timeless reality, which is "already all laid out" in *both* past and future realms. In sum they are known as Block Universe concepts, a term suggested by philosopher William James over a century ago.[1]

For example after about 1840, the smooth motion of a railway carriage, suggested comparison with the seemingly similar smooth passage of time. So that Zeno might well be using railway carriages for his paradoxes, were he alive today!

Imagine then you're a passenger on some metaphysical train progressing through the country of reality – and crucially seated with your back to the engine at all times. Your *present* is your impression of those trackside poles, which periodically loom up at your window as you move along. Your *past*, or memory, is your fading vision of them, as they recede downline.

This model also allows for occasional glimpses of the *future* as construed through anti-memory. For sometimes somehow you might succeed in stirring those mental muscles so long grown rigid by gazing mainly towards your *past*. Then you could change your time perspective to glimpse *future* poles further up the line, before they come into the *present* or abreast of you!

Another Block Universe model goes back to the great mediaeval philosopher St. Thomas Aquinas (1225-74). He wondered if, just as a pilgrim can know the road of life that stretched behind him, might God not know the rest of the road which lies ahead?.

This model can also be recast into modern city terms. Imagine therefore that all reality (*past, present, future*) is laid out in static form, like those blocks of buildings, laid out in grid plan for New York's avenues and streets. Along one of the latter you are progressing, somehow constrained by your physiology to the same fixed rate through life.

Imagine further some long-lost friend likewise proceeding along an avenue perpendicular to your own particular street. To some Godlike observer in a helicopter high above, it's perfectly clear that the two of you will meet soon. This meeting is fated for the next corner, where your two routes will intersect.

Normally of course neither of you could know in advance, about this unexpected meeting just ahead. Paranormally however, some act of intuition or anti-memory, might well have given one of you some foreknowledge of what is to come!

A more sophisticated Block Universe model came with the invention of moving cinema pictures ca. 1890. The cinema gives the illusion of flowing continuity, through phased illumination of slightly differing pictures sequenced on film. Your *present* is then provided by whatever section is currently under the flickering projector light. It has followed from all those *past* film slices now rolled up unreachably.

Your *future* too will likewise follow, from those film sectors not yet reeled out, and so equally unreachable!

This model can also be improved if we discard those two cumbersome film reels. Imagine instead that your entire lifetime history is laid out in one long static strip of film, each tiny picture with blank spaces in between.

Above this filmed record of your entire lifetime, your Mind is (somehow!) travelling at a fixed rate. It can view only what is directly below it, through some narrow stroboscopic flash of consciousness. This lights up about 20 times per second, avoiding blank spaces but illuminating each little static picture of your *present* as it moves along.

All of which confers the comfortable illusion (much as in a railway carriage) that you are somehow static - while life's scenes are clearly passing by. But of course in reality it's only your Mind which does the changing, forever passing at a fixed rate above that static record of your entire history!

This stroboscopic model is also easily enlarged, to include the experience of memory in both modes. For normal memory you need merely imagine that flashing light of *present* consciousness, expanded in focus to include *past* pictures in its beam. For paranormal anti-memory you can again expand the focus forward, to include those *future* pictures still to come!

So that again technology provides a new time model, and one which Zeno would likely have approved.

Enter Einstein

Block Universe thinking like this also received a most unexpected boost, when Albert Einstein (1879-1955) introduced his first or Special Theory of Relativity in 1905.[2] He started by noting that nothing in the entire Universe is ever really at rest.

For example you may think that you're seated comfortably at rest as you read through this book. But your chair is just part of the Earth whirling through space (at some 60,000 miles an hour!) relative to our local sun. Our sun in turn is whirling round relative to The Milky Way, our galaxy likewise is flying apart from all others in the Universe, and perhaps so on infinitely.

All of which means there can't really be any fixed framework, for Absolute Space as Newton had proposed. And because both are so intimately connected, his further postulate of Absolute Time then becomes equally untenable.

Einstein further noted that proper description must always include the *time*, at which any measurement is made. For example to properly describe some tree as 30 feet high youneed to know the time at which this measurement was made. For it was undoubtedly less high years ago when a seedling, and will again be less high when it falls down some day.. This is the significance of time as the familiar Fourth Dimension (i.e. measurement), the other three of course being length, breadth, height!

A further observation by Einstein was that nothing in Nature ever seems to travel faster than the speed of light. So he made this fact into another universal principle. This upper speed limit then becomes a natural constant, a sort of physical boundary throughout the universe

. Light's natural upper speed limit also turns out to be very important for time. This is because your timing of separated incidents must depend on clocks, which at best can only communicate at this upper-limit speed. From which it follows that your judgment of *precedence* – that apparent time sequence which you assign to any chain of events - depends entirely on your location in the Universe.

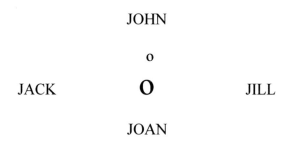

DIAG. 18.1: Dependent on their position in space, each observer will make different space judgements – *beside, before, behind* **– on how the golf ball is situated relative to the hole.**

With space people have always accepted precedence as a purely personal judgment, one with no real significance in the overall scheme of things. For example in Diagram 18.1, Jack and Jill are located to either side of a golfing green. Whence they must judge that the ball now lies *beside* the central hole. But John to the north will see that it's lying *be-*

OR DO WE PASS THROUGH TIME?

fore it, while Joan to the south regards it as obviously *behind* ! Still all are agreed that these are purely local space labels, ones entirely dependent on where each observer is. So that our various judgments of *beside, before, behind* have no real significance in the overall scheme of things.

With time however a similar personalization of precedence can be far harder to accept. Nevertheless consider explosions in two separated stars as in Diagram 18.2. The light emitted by each explosion is the only way we can know about such events. So Joan will then judge that Star A exploded long *before* B.

But John can only judge the converse - that A's destruction must have happened long *after* B. While Jack and Jill will think that both stars flared up s*imultaneously*. Your time labels therefore depend on how long it takes light to reach out, to wherever you happen to be situated in the Universe.

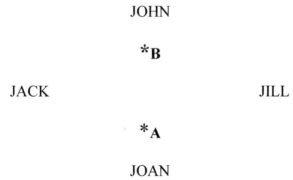

DIAG. 18.2: Since light takes time to travel from each of the two stars, the various observers will make different judgements of time precedence – earlier, later, simultaneous – about whether A or B exploded first.

In the small slow regions of everyday experience, we're normally unawares of such variability in precedence. But that's only because the speed of light is so high (300, 000, 000 metres per second), and the speed of thought in our crude ion-based senses 3 million times more slow. [3]

Imagine however that our four observers had much swifter electronic senses, with thoughts which could discriminate between light signals just 1 billionth of a second (10^{-9} sec.) apart. Light can only travel one foot of distance in that brief interval.

Imagine further that you have replaced the two stars in our diagram by two duellists at 50 paces armed with their pistols. While Jack and Jill might see them firing simultaneously, you could have endless argument between John and Joan about who really fired first!

All of which means that strictly speaking, the instant of Now cannot extend beyond the point of Here. In fact when you sit across the table just three feet from your partner, your respective Nows are about three-billionths of a second apart! Though of course our senses are far too slow in function to appreciate such tiny intervals!

This further means that Newton's prime postulate of Absolute Time, or a universal Now which extends across all the Universe, cannot be true. Instead there's an infinite number of different instants in time which may be termed Now, an infinite number of different points in space we can call Here. Likewise there's an infinite number of personal futures, and what people term 'the future' is really an amalgam of them all.

Minkowski and space-time

Algebra was the mathematical method through which Albert Einstein expressed all this in his first Theory of Relativity (RT), 1905. Three years later his former teacher, Hermann Minkowski (1864-1909), recast this algebra into new diagrams of geometry. This simplified Einstein's treatment, allowed peoples' visual intelligence to come into play, and so made RT much easier to understand.

In a famous speech at the University of Cologne – on Sept.21, 1908 - Minkowski began with a strictly factual observation, very much as Einstein had done previously:

> "Nobody has ever noticed a place except at a time, or a time except at a place... Henceforth space by itself, and time by itself, are doomed to fade away into mere shadows, and only a kind of union of the two will preserve an independent reality."

Minkowski's treatment therefore saw time and space as conjoined together to constitute one reality, where previously they'd been viewed as separate aspects. So a point in space at some point in time becomes a *world-point,* a particle which endures in space will have its own *world-line,* and any intersection of two such world-lines becomes an *event.*

The entire Universe then consists of the totality of such events and world-lines – all lying static and displayed in the new reality of *space-time*:

> "the whole universe is seen to resolve into similar world-lines - and physical laws might find their expression as reciprocal relations between these world-lines"

OR DO WE PASS THROUGH TIME?

When space-time is simplified or slimmed down, from 3 to 2 space dimensions along with the time dimension, it can be illustrated as a diagram with two opposed Minkowski cones. These cones meet at a very special point described by the observer as Here-Now. Through it runs the vertical axis of time, which is the observer's personal world-line.

This personal world-line in the Minkowski diagram can also be termed your *life-line,* one which details your entire history from womb to tomb. As described by philosopher Arthur Eddington (1882-1944), it resembles some long, narrow, rubber worm - one which lies tenseless and static in space-time. [4]

In the 3 space dimensions your life-line's greatest extension would be about 2 metres, assuming that you're over 6 feet tall. But in the 4th dimension of time its extension might be 3.2 billion seconds (3.2×10^9) at most, that is if you live long enough to become a centenarian!

Other world-lines intersect your own personal world-line as *events* you observe at varying intervals. If straight, these other world-lines depict people or objects moving at a regular speed. If curved they depict acceleration towards or away from you.

It's easy enough to see how events in your personal *Past* might be described in space-time, through tenseless terms of other world-lines intersecting with your own. But the problem is that *Future* event-intersections appear to lie equally tenseless - and "already written" on the space-time diagram as well!

This is because, as we've already seen, our deeply cherished judgments of past and future are purely personal and local labels, but of no ultimate significance in physical reality. Which also makes the space-time treatment essentially tenseless, with time reduced to a sort of space in the mathematical formulae.

Different observers, each travelling at their own velocity, will therefore always slice up reality into different planes. In the old world of Newton, reality would be like a pre-sliced loaf, each new slice denoting another instant of Now across the Universe. [5]

But in the new world of Einstein-Minkowski, reality is more like a solid loaf which can always be sliced through in all directions, so giving an infinity of different planes. Each of these planes will have differing allocations for space and time. Space is then dynamised, or time spatialised, in different degrees. It all depends on the orientation of one's personal world-line.

As partly described in Diagram 18.2 also, observers with different velocities, must always observe time precedence differently. So that an event labelled *Past* by one observer may well be labelled *Present* by another, or even *Future* by a third. For as we've already seen, these purely personal time labels have no real or absolute significance. In this they resemble those equally personal space labels of *'behind, beside, before'*.

All of which led Einstein in his last years to remark that the common distinction between past and future is "*only an illusion, although a persistent one.*"

How real is space-time?

Throughout basic physics Einstein has therefore totally superseded Newton, though the latter is still reliable enough for most everyday affairs. Otherwise RT is now widely used as a routine mathematical tool in physics, repeatedly verified as totally reliable and dependable

For example atomic energy came from its predictions over 60 years ago. Nowadays also Global Position Receivers – which may tell where your car is located on a roadway - can only function through RT mathematics. These calculate your proper time as different from the common or Newtonian variety.

There is however one very important sector, where RT's apparent implications have never been resolved. This is our purely personal experience in what Einstein termed 'middle-time' – that everyday region of days, hours, minutes, seconds against which all our lives are run. Here RT's apparent consequences - or what it all means in reality for the human condition – is still quite totally unclear.

RT is similar to quantum theory in this respect. Both these marvellously accurate physical descriptions seem to imply consequences apparently at variance with everyday experience.

For RT the problem is that it seems to conflict directly, with all our normal temporal ideas. Or as physicist Paul Davies says: [6]

"The greatest outstanding riddle concerns the glaring mismatch, between physical and subjective, or psychological, time"

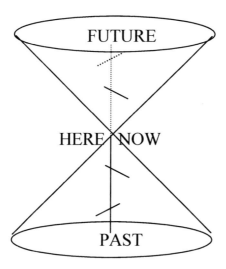

DIAG. 18.3: Einstein's theory of relativity can be expressed as space-time diagrams bounded by Minkowski light-cones. Events in your past can be depicted by the intersection of other world-lines with your own time-line.

Events in your future may be apparently "already written" likewise, although not normally known to you now.

Davies terms this mismatch *"Einstein's unfinished revolution"* – and it's just about the only practical problem, which RT has ever encountered so far. But then again – with the single exception of new anti-memory experiments I'll report in Ch. 19 - it's never been probed pragmatically at all....

Apparently Minkowski himself regarded his new geometry of space-time as more than just another mathematical tool. Rather he saw it further as a true ontology, or accurate description of the natural world. If so, the latter must be essentially static in reality. In which case space-time (or strictly most commentators' interpretations of it!) seems to support that old philosophy of *Being*, remarkably much as Zeno advocated 2,400 years ago!

What then can we make of our everyday impressions of *Becoming* – that ceaseless experience of change and motion as allegorised by Heraclitus, and which so dominates our lives?

As I've noted earlier, it's quite easy to see how all *past* events in your lifetime, might be depicted in the lower half of your own personal Minkowski diagram. The problem lies with the upper half. For what you see as the *'not yet future'* is merely a local and very personal label, one which may well be regarded as *'already past'* by some other observer's different world-line!

Apparently therefore space-time requires that all those events which you term *Future* must be seen as "already written" – and so of very similar status to those which you term *Past*. And since both these purely personal or local labels are of no great significance overall, the totality of all events is best described without these labels, i.e. in *tenseless* form. If so, events may not really "happen" but merely "are". So that we may just "come across them" as our lives proceed!

In 1922 logician Hermann Weyl (1885-1955) was one of the first to express this aspect unequivocally: [7]

> "The scene of action of reality is not a three-dimensional Euclidean space, but rather a four-dimensional world in which space and time are linked together indissolubly......
>
> "Only the consciousness that passes on in one portion of this world experiences the detached portion that comes to meet it, and passes behind it as history, that is as a process that is going forward in time and takes place in space."

So too said French physicist Costa de Beauregaard in 1968: [8]

> "Humans and other living creatures, for reasons which one can try to explain, are compelled to explore little by little the continent of the fourth dimension, as each one traverses, without stopping or turning back, a time-like trajectory in space-time."

In 1974 – a year after I first publicised the pre-call concept - physicist Gerald Feinberg commented on this aspect. He noted, that since the equations of relativity involved squared functions, their roots can be interpreted in advanced (+) or retarded (-) terms. [9] This is another expression of those advanced and retarded functions we've already encountered in Ch. 16.

The retarded or (-) solution may then be correlated with the bottom or non-problematical Minkowski light cone. This is the normal one in which it's easy to imagine all your *past* life inscribed. So much is accepted because it has been so universally observed.

But Feinberg stressed further that the advanced or future solution – the top cone in the diagram – has only been disregarded because it's never observed in reality. Though 'never' might well be replaced by 'seldom' - if what he termed 'precognition' (sic) were ever firmly proved.

Feinberg's suggestion can now also be recast in much firmer anti-memory terms. That this new paranormal development is likely to be found of general application now seems probable. If so, its validation would suggest that the top of Future half of the Minkowski diagram can also be filled in at least partially. Though just how far must require much more experiment to decide....

Philosophers too have tended to accept the reality of space-time. One of the most forthright was Desmond Williams in a famous article on "The Myth of Passage": [10]

> "I believe that the universe consists, without residue, of the spread of events in space-time,...
>
> "the theory of the manifold is anyway literally true and adequate to that world: true, in that the world contains no less than the manifold; adequate, in that it contains no more."

So that as Paul Davies again maintains: [11]

> "Relativity has shifted the moving present out of the superstructure of the universe, into the minds of humans where it belongs."

What Davies seems to imply here is that we may require some new way of thinking about time, which doesn't just assume that it passes by. And from this new viewpoint seek to explain why everyone should imagine that it does!

Though here a purist might observe that perhaps it's not really RT, but rather Davies' interpretation of it, which has shifted the moving present from the universe!

Similar themes are echoed by leading British physicist Sir Roger Penrose. In *The Emperor's New Mind* (1989), and also *Shadows on the Mind* (1994), he notes that RT has always proven marvellously accurate wherever it's been tried. So how to explain its large apparent conflict with that common but untested notion, that time is somehow flowing past ourselves? [12]

> "Consciousness is after all the one phenomenon that we know of, according to which time needs to 'flow' at all.... My guess is that there is something illusory here too, and the time of our perceptions does not really flow in quite the linear forward-moving way that we perceive it."

Elsewhere Penrose queries: [13]

> "If the equations of physics seem to make no distinction between future and past – and if even the very idea of the 'present' fits so uncomfortably with relativity – then where in heaven do we look, to find physical laws more in accordance with what we seem to perceive of the world?"

To which the new anti-memory outlook can now afford one clear reply. It's not the well proven laws of physics that need changing for better harmony, but rather our common perceptions of the world. For those laws of physics have been tested to exhaustion and never found wanting – whereas common notions about time 'passing' have never been tested in any way at all!

And then of course there's *The Prime Assumption,* along with those various anti-memory observations that can now so readily falsify the same…

In his famous best-seller, *A Brief History of Time,* cosmologist Stephen Hawking has also pondered the problem of memory's time asymmetry: [14]

> "Why do we remember the past and not the future? It is rather difficult to talk about human memory because we don't know how the brain works in detail…"

This question he then proceeds to answer, in terms of entropy. Linguistic analysis however suggests a rather simpler conclusion here. For since 'future' implies 'not yet', while 're' means 'again or afterwards', Hawking's question just reflects confusion through oxymoron.

His entire question then seems better reframed in less self-contradictory terms. For example had Hawking queried *"Why can't we pre-call as well as re-call?"* he might well have keyed into more productive territory!

In sum in any case likely most physicists and philosophers now accept the concept of static space-time, as a valid description of reality. The striking reliability of RT, and its total verification wherever it's been tested, have enforced this view.

However as we've previously considered in the context of specialisation or *Inverse Expertise*, physicists are unlikely to be too well versed, or even interested, in psychology. And certainly not in any paranormal possibilities like intuition, pre-call, anti-memory!

Though of course whether Mind or physics is more basic might be quite difficult to decide….

So far however in any case, nobody has ever explained how the physical picture of space-time, might be reconciled to the more psychological realms of ordinary experience. Neither has anyone ever suggested any experiments whatsoever, which might help cross this great divide. On all of which therefore anti-memory now has some relevant things to say....

A brief summary

Common thinking still supports the old time philosophy of *Becoming*, as expressed in the notion that "Time (somehow!) passes by". However the rival philosophy of *Being* has never quite gone away, It holds that outer reality is in essence static, while we are (somehow!) "just passing through."

Being gained much apparent support, from Einstein's introduction of Relativity Theory (RT) into physics a century ago. Its concept of space-time shows that *future* and *past* are merely personal time labels. So that events overall (i.e. in both personal past and future) seem best regarded as "already written" - while we merely "come across them" as our lives proceed.

The first problem is therefore how RT's apparent world-picture, could ever be tested out against everyday experience. And the second problem would be, if such an experiment proved positive, how to reconcile its findings, with our everyday impressions of reality...

19

TESTING FOR RELATIVITY

"It is my personal belief that we are approaching a pivotal moment in history, when our knowledge of time is about to take another great leap forward."
Paul Davies - *About Time* – 1995 – p. 283

IN THIS SEEMING CONFLICT between common impressions and RT - or between *Becoming* and *Being* - the pragmatism seems all on Einstein's side. For RT is a theory, which has been tested to exhaustion, and always emerged triumphant everywhere. Whereas that common impressions that "Time is (somehow!) passing by us" has never been tested, even slightly, in any form at all!

These common impressions are also of course entirely dependent on that *Prime Assumption* I've already clarified. For people have merely *assumed* - but without any testing whatsoever - that Mind or memory can just directly contemplate the *past* alone. An assumption which the continuing reality of intuitions worldwide, together with all the new anti-memory findings, now indicts as suspect and quite likely wrong.

New fit for mind

Objectively therefore, and regardless of subjective feelings or even unrealised censorship, it must seem best to adopt an RT viewpoint in these affairs. Which then leads us to consider common or everyday time, not from the viewpoint of traditional beliefs, but rather in 4-dimensional or space-time terms.

And when this RT viewpoint is adopted, one new psychological generalisation or summary immediately results. It's a summary of somewhat compelling symmetry, indeed so much so that that one might almost wonder why anti-memory should not be!

Let's suppose therefore that life really is conducted, in that 4-dimensional reality which RT implies. Taking first those 3 familiar space dimensions (i.e. length, breadth, height) it's clear that Mind or consciousness views all of their 6 possible directions equally. That is to say we're aware to the east-west, north-south, up-down – and all in a wholly similar way.

In addition this awareness, by an Observer at Here, shades off with space intervals (i.e. distance) exponentially. We can know ever less about distant presentations, the further they're away. This follows from the inverse square law of physics, which affords an exponential curve of steeply decreasing information about far-off objects, as their distance from an observer grows.

But in terms of physiology or personal impact, this exponential curve is flattened out, or rendered much less steep by the well-known Weber-Fechner law. This states that increasing increments of sensation, are only registered logarithmically by your brain.[1]

For example suppose that some noisy neighbour in the next room turns up the volume on his stereo by a physical factor of 100 or so. In that case your physiological or personal increment of annoyance will only increase by the log of 100 which is just 2!

Turning next to the 4th dimension of time with its 2 possible directions, we're aware of the *past* in a broadly similar way. For we tend to know ever less, about personal past experiences, the further they recede in time. This is a matter first clarified by H.Ebbinghaus, whom I've now cited repeatedly.[2]

All of which then means that as events grow ever more distant in space or time-past, and in terms of information available at Here-Now, the psychological labels for space and time-past are largely interchangeable. This novel unification can also be summed up in one simple expression of 4-dimensional symmetry, as first published by me in 1973:[3]

> Mind has a 7/8 symmetry of outlook, in the 4-dimensional reality of space-time.

And of course when the matter is put like this, one obvious question immediately suggests itself:

> Why not a full 8/8 symmetry - to confer complete awareness, over the 8-directional or 4-dimensional world?

Theoretical treatment of full symmetries in this way can often be very rewarding, as physicists in particular well know.[4] Considering therefore Mind's 7/8 symmetry of outlook on 4-dimensional reality, intuitions of course suggest that the missing 8th element of future awareness, does in fact sometimes occur. They appear to happen with a relative frequency of about 1 mentality in 10 million (10^{-7}), as I've already shown. (Ch.4)..

Further, as I've reported throughout Part Three, this latent pre-call or anti-memory faculty can be developed readily. Though so far these developments I've reported (cards, roulette, etc.), have been really rather limited in both space and time.

But in any case this novel anti-memory possibility now proffers a new link or harmony, between physics and psychology. Like a glove on the hand, it can unite these two formerly disparate disciplines, in a manner impossible before. Even though the physical conversions between time and space are complex or non-Euclidean, whereas the psychological conversions involve a simpler form of symmetry.

Such a new unification, between psychology and physics, is properly termed an *isomorph* (lit: *same shape*). It's a requirement, for which time scholars like J. Piaget or J. T. Fraser, have expressed the need.[5] It's also what Jung the psychologist and Pauli the physicist sought vainly through Synchronicity (Ch.8).

But now this long sought ideal can be achieved quite readily, through intuition properly understood. It emerges when anti-memory fills out full psychological 4-dimensional awareness of reality. With just one simple postulate it therefore suggests where conventional thinking has been lacking all along. And where a fuller picture would resolve the seeming conflict with space-time.[6]

Anti-memory further fulfils the requirement for some new outlook on time, as requested by J.L.Synge long ago, and other physicists more recently.[7] Finally it clarifies where previous logic has been lacking, through that totally untested *Prime Assumption* that memory is restricted to the past alone.

All in all therefore it now seemed to me, that there were so many good and objective reasons in favour of anti-memory, that any true believer in RT might almost think that it should be! So this was the mode in which I first expressed the concept, at London in 1973.[3]

Further of course nobody until then (nor indeed since!) ever seems to have ventured to test out RT, as it may apply to everyday reality and Mind. A test which anti-memory should enable quite readily, and which I therefore decided next to do!

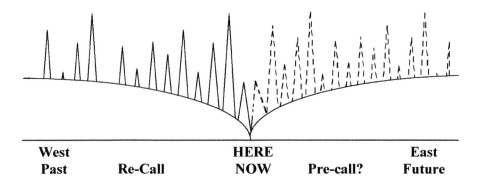

DIAG. 19.1: Knowledge attainable at HERE-NOW decreases exponentially with distance for all 6 possible directions in space (West-east. North-south, Up-down). This further applies with duration to the 7th direction of time-past, so that a 7/8 symmetry of outlook results.

Full future-oriented pre-call or anti-memory capability would fill out this 7/8 symmetry totally. A new isomorph or parallelism, between physics and psychology, would then result.

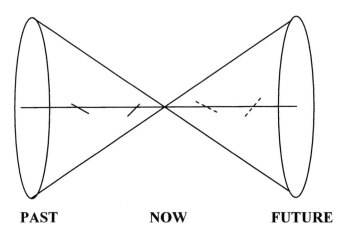

DIAG. 19.2: The top diagram also bears useful comparison with Minkowski space-time treatment (time axis horizontal.) Anti-memory treats your Future event-intersections as perhaps knowable, and so similar to your Past ones - as RT also apparently implies.

People with numbers

This was the focus of my fifth new experiment with time. And it proved by far the most wondrous and astonishing of all up until then. So that those events and developments which I have to relate next, were indeed most striking, curious, extraordinary.

In fact they may seem so extraordinary and 'far-out' – both in their methodology and content - as to invite almost automatic rejection through unrealized censorship from the overly-sceptical.

Nevertheless my findings, in this fifth and final new time experiment that I will next report, were objectively all more or less as RT would imply. The striking thing was that they seemed to fulfil its apparent consequences so faithfully…

What I'd been seeking at this point was some new and far wider pre-call system or routine - but one still wholly numerate. It should enable numbered testing of anti-memory against the fabric of everyday experience. It would be applicable to ordinary people, as they went on living normally in their everyday routines.

It was while pondering this requirement at a very early stage, that I suddenly realised that there is one everyday activity, which is indeed quite numerate. This happens whenever people are driving motor vehicles. For then they're compelled by law to carry numbers around with them - in the form of their licence-plates! These have digits which are largely random - and purposely designed for all to see!

Considered like this, all passing traffic reduces to a stream of people (i.e. the drivers) all labeled by random numbers - and all highly mobile over large intervals of local space and time. Ordinary traffic therefore provides a random number stream which is nearly perfect, and so ideal for pre-call experiment.

The main imperfection is that there will be slightly more low numbers than high ones. Other slight imperfections – for example local buses with sequential numbers – are easily rinsed out by protocol. (Ignore all buses as they come along!)

Car-spotting!

I termed this new pre-call mode of mine 'car-spotting', in the interests of descriptive accuracy.[8] It would focus on vehicle number-plates, in a wide variety of situations and routines. Always too it would be entirely and completely numerate – so that exact calculations of chance probability could be achieved.

Car-spotting can also readily be carried out, over many miles and hours of local space and time – i.e. all over the local sector of space-time! Which further explains why it soon emerged as quite the most interesting and informative exercise of all!

Soon therefore I started to pre-call car number-plates according to those few conventions described below. And over the years, and many thousands of trials, I found that my competence worked equally in Ireland, Britain, Germany, Libya, America. So that likely it would also function anywhere.

More recently in Britain however, licence-plates have been modified away from random 3-digit numbers over the last few years. But otherwise I never found any real difference in competence between these various territories worldwide.

For this novel exercise I also transferred those 3-digit conventions already described (Ch. 13.) This meant I would focus only on *the last 3 digits* on any licence plate. For example if some number like 4 **7 6 3** came along, only **7 6 3** would be relevant to my scheme. Conversely if some single number like 6 happened by, that would go down as **0 0 6** in my records.

Those attitudes which I brought to this new pre-call exercise, were also all very much as before. They merely involved a deliberate mental calming, maximising that Childlike state I have described before. Then I would try to focus clearly on whatever 3-digit number that I might next see. Again this was all done, more or less as I've previously described.

Soon then I was gratified, not to say astonished, to find that this fifth major pre-call learning routine was the most revealing and informative of all. Not only that, it was by far the easiest anti-memory mode as well!

That car-spotting proved out as by far the easiest anti-memory exercise, is now of course readily explained. For licence-plate numbers always manifest amid huge variety. They come associated with a great range of different shapes, noises, colours, speeds. As such they're inherently *more interesting,* than dull lists of random numbers displayed on paper or computer screens

Car numbers are therefore more likely to appeal to the Child in one's personality, as a new and exciting sort of intuition game. Or in terms of the basic pre-call diagram (Ch.4), they're more likely to project quite strongly, above either memory threshold.

Extending into time

Firstly therefore I started this exercise with a pen and notebook, parked in my car by the side of a suitably quiet road. Then I would try to pre-call whatever 3-digit number might next appear, on the next vehicle passing from behind. Those roads best suited to this purpose – neither too busy nor too quiet - would afford a decent experimental rate of 50 to 100 trials per hour.

In any case, after some 100 hours or 10,000 actual trials, I found I could easily maintain routine low-level 3-digit scoring, at very high levels of reliability. For example unplaced single-digit scoring (X-type) would then normally be around 90%, unplaced double-digit scoring (XX-type) around 50%, and triple-digit (XXX-type) around 5% capability.

As previously also, low single-digit scoring rather felt like being able to hit a metaphysical 3-digit dartboard at some place, and with even careless effort, nearly all the time. Though it still required more steadiness to get higher double-digit scoring, and almost superhuman effort - at first - to achieve ideal 3-digit scores.

And however one might chose to calculate statistically, the odds-against-chance were still far more than 1 million to 1 (10^{-6}), for even 2 hours with such competence.

Eventually I conducted numerous thousands of car-spotting trials in this way. Of these we need only consider 4 brief routine series here - part of an effort designed to probe the edge of the anti-memory envelope, the far outer limits of my skill.

Of these the most simple situation is as I've just described above – i.e. pre-calling the number of the next vehicle that would pass by. Here the time interval – between prior pre-call and later observation – was something like 30 seconds on average. To this extent I was aiming to go into the future of my own awareness, each time I made a new pre-call attempt.

Secondly I then extended this simple exercise, much further into future time. Here my procedure was to set a pocket alarm to ring, at some fixed time like 15 minutes ahead. Immediately after setting the time I would try to pre-call the relevant number. Then I would wait to inspect the first vehicle, which would pass by after the alarm had rung!

To avoid confusion in both modes, it's best to wait until the road has fully cleared. My target was then always specified as the first vehicle to come along *after* this was so. Naturally also that 15-minute interval makes trials far slower – and ever more so with further extension into future time.

In this way in any case I gradually extended the time interval up to 10 hours ahead. Sometimes this meant pre-calling at night, for an observation to be made the following morning! And yet, to my general surprise and as Table 19.1 again shows, I found little or no diminution in pre-call competence - even at such very long intervals.

That deliberate numerate pre-call can function, over at least many hours into the future, was therefore quite clear. There appeared to be no obvious edge to the anti-memory envelope, even after such long intervals. Though of course a more general analysis, of spontaneous intuitions in real life, would also suggest the same.

Extending into space
Thirdly I reverted towards a more space-oriented exercise. Here my aim was to pre-call the number of some vehicle parked round a suitably blind corner, initially perhaps 200 yards away

And just as before I had extended into time, I now gradually extended into space as well. So eventually I ended up pre-calling parked vehicles up to 100 miles away. This was a practical limit imposed by happenstance ordinary car journeys undertaken for other purposes.

But again anti-memory functioned here very much as usual - with 89% X and 50% XX scores. Also I found surprisingly little reduction in competence as the distance grew.

To pre-call some car parked at a distance in this way would have been attributed to 'clairvoyance' by the mesmerists (Ch.5), or likewise to 'remote-viewing' in more recent terms (Ch.9). But here anti-memory was always demonstrably a more logical concept. For I was really targeting whatever number I would find *whenever I arrived* at that distant place. And over some hours this was in general unlikely, to be the vehicle parked there when my initial prediction was made

But in any case it was now clear that pre-call could function over many hours of time and many miles of space. It was all very much as would be required, by full 8/8 symmetry over local space-time!

An appointment with Fate?
Finally I combined time and distance to produce a closer psychological version of space-time. Here I would record my pre-call statement (or 'prediction'), before setting out on some local journey of 10 miles or so.

Once more when my pocket alarm sounded, after a typical 15 minutes of time and wherever I happened to be in local space, I would stop (as soon as the traffic permitted!), and wait for the road ahead to clear. Only then would I record the first number to come travelling towards me. In this way I was therefore combining local space and time.

Yet still anti-memory continued to function very much as usual, with 92% X-scores and 50% XX-scores in this mode too.

SCORES	UNPLACED			PLACED		
	X	XX	XXX	+	++	+++
Chance/100	(65.7)	(15)	(0.6)	(30)	(2.8)	(0.1)
1/ TIME						
30 secs. ahead	94	42	1	48	9	0
2/ TIME						
To 10 hrs ahead	96	38	1	42	9	1
3/ SPACE						
To 100 miles away	89	50	9	41	14	2
4/ SPACE+TIME	92	50	3	53	11	1
TOTAL/400	361	180	14	184	53	4
CHANCE/400	(263)	(60)	(2.4)	(120)	(11)	(0.4)

TABLE 19.3: Though car-spotting in local space and time was carried out under 4 very different conditions, overall pre-call competence remained much the same throughout.

Meantime the whole flavour, of car-spotting in these last more extended modes, is best encapsulated by one especially memorable incident, one not included in the totals above:

> Before driving southwest to Galway, from my parents' home in Ballybay, County Monaghan, I decided to test out pre-call on some dozen vehicles, we would meet at fixed points about ten miles apart.
>
> These points were chosen as those standard traffic signs, outside each village on the trip.
>
> The first half-dozen results proved quite significant as I drove along with my sister Anne: the expected total for double-digit (XX) scores was building up quite well.
>
> Our next observation point was outside the village of Ballymahon, for which I'd pre-called a 3-digit number over 2 hours before.
>
> At 2.18 in the afternoon we stopped there. As usual we waited for the road to clear, then watched for my target to come into view. It proved to be an old black Ford Prefect, whose slightly wavering progress suggested a local farmer driving home 'happy' after the fair.

But his 3-digit number was 4 8 5 – exactly the same as I'd pre-called some 70 miles back and over 2 hours before!

Was this merry farmer then really subject to predestination, as he worked his way homewards through the morning fair? And if so were we not subject to the same fate also, being destined to meet him at this pre-called place and time?

I decided to leave such weighty implications to others who might care to ponder them if so inclined. For myself it seemed far wiser not to dwell on them too much at all. Much better just to keep on working, and consider my results as they now accumulated rapidly!

Little by little I therefore learned to curb my initial astonishment, and even awe, at such 'far-ahead' pre-call feats. Slowly thereafter they became a matter of routine. And from there they evolved into just another interesting and challenging intellectual exercise.

Now too I found other accumulating analogies with normal or past-oriented memory. For example, there's nothing at all unusual in being able to *re-call* the number of some vehicle you encountered 70 miles, or 2 hours, back along the road. So to find that I could also *pre-call* a similar event – from 70 miles or 2 hours further ahead down the same road - then conferred a certain degree of elegance in personal time symmetry. And of course it all implied a new and better fit, between everyday experience and the RT description of space-time.

Choices predestined [9]

While writing this book I was long gone out of practice, in my old 3-digit car-spotting routines. But now I decided to re-test the matter, and also find out how well I could resurrect my former competence. Nor was I at all surprised to discover that it took me a good 10 hours of retraining, before I could regain a decent semblance of my former skill.

After which I decided to experiment once more with car-spotting, though in a slightly different mode. This time I would more definitely investigate the role of Fate and free will in the lives of those concerned.

I would therefore attempt pre-call the next vehicle to stop at a pedestrian controlled crossing. This future event would at all times be controlled by whatever casual strangers happened to press the button to get across the road. As usual too some minor traffic conventions are required here, for example to focus on just the nearside lane in a multi-lane street.

I conducted the first such numerate "Experiment with Fate" in Central London, with a total of 200 trials. The entire effort therefore involved 8 series of 25 individual trials apiece, conducted during early August,

2004. Each of these series required about 1 hour to complete. My location was on Euston Road, beside the main gate of the British Library, where incidentally much of this book was researched.

There's a controlled pedestrian crossing at this location, and also a low white wall where you can conveniently sit down. Here the pedestrian button doesn't operate very quickly, because of the high-volume traffic on the road. Additionally during August 2004, only about one-third of the passing London traffic still carried the old 3-digit number plates.

Nevertheless, after settling on some minor traffic conventions, it proved relatively simple to conduct the standard block of 25 pre-call trials as usual. The combined results are given in Table 1 below. Effectively they return a diminished version of those scoring levels I had attained of old.

Still my conclusion was quite unavoidable and clear. At least this little local slice of London's future, was all more predetermined, than either driver or pedestrian would ever realise. Something resembling the popular notion of Fate was therefore operating in these cases, over about two minutes into future time. But again of course such was always implicit, in those Minkowski space-time diagrams.

My second such experiment I conducted in Galway, Ireland - going back to those familiar but now much altered scenes where all this had started decades before. Here my location was outside the estimable Maggie's Café, down near the historic Spanish Arch. Nearby there's a push-button pedestrian crossing, one operated very frequently by locals and tourists alike. My standard 25 trials therefore took only ca. 40 minutes. As before I conducted just 8 of these in all.

My scoring was well down on this occasion, again easily explained by lack of practice over recent months. But still it was sufficiently significant. That the good citizens of Galway, were then just as susceptible to pre-call as their London counterparts, was in any case quite clear. Nor indeed did this surprise me very much, considering that the whole idea had emanated from there many years before....

To successfully pre-call a significant portion, of whatever number will halt next in your vicinity, and at the behest of some pedestrian supposedly operating under free will, can prove a very sobering experience indeed. As before it inevitably implies, that the future of these people, is to some extent 'already written' all unknown to them.

This is of course as RT apparently requires. But there's also the further unavoidable implication - that your own fate must also be predetermined to a similar degree.

	UNPLACED			PLACED		
	X	XX	XXX	+	++	+++
CHANCE/100	(65.7)	(15)	(0.6)	(30)	(2.8)	(0.1)
Euston Rd.						
1/	97	52	4	56	7	0
2/	96	41	2	43	9	0
Spanish Arch.						
3/	81	20	2	32	5	1
4/	78	35	2	39	9	0
TOTAL/400	352	148	10	170	30	1
CHANCE Exp:	(263)	(60)	(2)	(120)	(11)	(0.4)

TABLE 19.4 : 3-digit observations at pedestrian crossings again proved amenable to pre-call, suggesting that their button-pushing choices were predetermined to an unrealised degree.

At this weighty point therefore I'll end my account, of this fifth major anti-memory experiment with time. Most of those conclusions I'd reached here were confirmed partially by others, and largely by my colleague Frank Morgan in Atlanta. Nothing whatsoever suggests that the same findings should not be broadly replicable, by anyone sensible so inclined. Rather their whole tenor is that more or less anyone can do the same.

For those philosophers and physicists who have pondered the problem of space-time as reality, the road to more pragmatic progress therefore now seems opened up and clarified. A few hundred active hours devoted to pre-call experiment, should produce more hard factual conclusions, than centuries of speculation ever did. Assuming of course that people can adapt successfully, to the considerable level of new thought required!

Likewise for all rightful sceptics who have long doubted the reality of the paranormal, the answer to their doubting now seems quite open and clear. They need merely take up some of those pre-call learning routines

I've furnished - preferably this last novel test for RT - to test out the matter for themselves.

If such sceptics then find they can't get anti-memory to function as I've described, they should publish their findings accordingly. Conversely if they do get it to function, most interest must attach to how much they may - or may not - observe those various secondary findings, which I've described along the way.

As customary with the rest of science, people can therefore now always freely avail, of that ever open Galilean invitation to "Try it and See!"....

A brief summary

If one applies RT and space-time thinking to normal consciousness, it suggests that people exhibit 7/8 symmetry of awareness, over 4-dimensional reality. Further consideration, of intuitions interpreted as anti-memory, suggests that the missing 8th element of future awareness might or should also be filled in.

This would confer full 8/8 symmetry of outlook over 4-dimensional reality. And so afford a long missing union between physics and psychology.

Still my first four pre-call routines were all rather limited, both temporally and spatially. But my fifth major innovation of car-spotting is near infinite in scope. It permits numerate observation of people, engaged in their everyday activities, across much larger swathes of local space and time.

These everyday futures can then be pre-called over extensive space and time intervals. The unavoidable conclusion is, that at least some of our personal futures in everyday experience, must be more predetermined than previously realised. But also of course as apparently required by RT and its expression as space-time....

To this extent at least therefore, ordinary reality seems all very much as deduced by quantum pioneer Louis de Broglie (1892-1987) long ago:

> Each observer, as his time passes, discovers so to speak new slices of space-time which appear to him as successive aspects, though in reality the ensemble of events exists prior to his knowledge of them.

de Broglie of course was expressing his purely mathematical or theoretical conclusions here. But the advance now is, that we can readily observe and investigate these things more pragmatically, i.e. through direct practical experiment…

20

A NEW KIND OF CONSCIOUSNESS...

> *"Scientists often miss things that are 'staring us in the face' because they do not enter into our conception of what might be true, or, alternately, because of a mistaken belief that they could not be true."*
> Peter Medawar (*Pluto's Republic* –1982)

MY SEARCH FOR THE MEANING of coincidence in every-day affairs had begun as just an intellectual pastime – little more than a novel challenge to intellect, and one mainly motivated by sheer curiosity. And yet, through strict application of scientific method, it soon afforded new unity where intuitions were concerned. Thereafter it led to new mental capabilities in several modes, ones that raised questions of some large concern.

That my new concept of anti-memory, is at least a viable new paradigm, or scientific hypothesis, then seems quite clear. For it can *explain* paranormal awareness much better than anything before.

To *explain* means to clarify what was formerly mysterious, in terms of something we already know. And that is an ideal which anti-memory now achieves.

Additionally however this new concept of anti-memory - or repressed pre-call to give it full title - can unite and integrate a great many facts, that seemed disunited and unrelated previously. Further it generates various new predictions. And a wide, open-ended, range of substantiating experiments which were previously unthinkable. The question is therefore whether this anti-memory concept represents a valid new theory – instead of just an interesting hypothesis?

What good theories do

Just what constitutes good scientific theory, has also been clarified by Murray Gell-Mann, famed Nobel particle physicist. Understandably he's been a stern critic of the paranormal as previously construed. But, in his book *The Quark and the Jaguar* (1994), he very helpfully summarises what good theories are and do.[1]

Gell-Mann lists out aspects of theory on which most scientists would probably agree. So a good theory allows for facts that were neither understood nor integrated previously. It will extend, and often alter, current knowledge. Sometimes too it can achieve a remarkable synthesis. It may compress newly realised regularities into one brief statement, regularities which have previously escaped notice in a whole range of phenomena.

In doing so good theory may also have to correct certain errors of interpretation accepted previously. The first details of its discovery may be somewhat messy, unsatisfactory, incomplete. But a really great theory can still unify with striking simplicity.

Elaborating on Gell-Mann, we can further note that good theories exhibit *utility* - because they "make more sense" of former mysteries. They may also enable increased *productivity* – in suggesting novel experiments inconceivable before. If so they should further afford reliable *predictability* - about the likely outcome of these new experiments.

A strong theory will also exhibit large *generality* – i.e. application to a wide range of phenomena. Finally it may incorporate *elegance* through concision and perhaps even symmetry. And if it exhibits all these qualities, people may well come to regard it as a valid expression of the natural world.

Good theories also help scientists because they don't have to think so much as formerly. No longer need they waste so much mental effort, in puzzling over what all seemed so mysterious before. Or as science historian T.E. Kuhn puts this aspect: [2]

> "Discovery commences with the awareness of anomaly... It then continues with a more or less extended exploration of the area of anomaly. And it closes only when the paradigm theory has been adjusted so that the anomalous becomes the expected"

All of which also of course could almost have been written, with the new anti-memory concept in mind!

The inference must then be that this novel concept constitutes a valid new theory rather than just a mere hypothesis. So in this chapter I'll consider some of those aspects in which anti-memory seems viable, and the new insights it can afford. After which I'll highlight some of those new problems it now raises, together with some of those many aspects still unclear.

An assumption falsified

Most importantly first, the new anti-memory idea directly challenges, and then negates through pragmatic experiment, that great universal

Prime Assumption which I've first stated in Ch.4. There I noted that people have always just *assumed* - but without any further consideration whatsoever - that Mind or memory can just reflect the past alone.

This is a belief so innate and apparently obvious, that nobody ever seems to have articulated it before. Much less proceeding to investigate whether it is really true!

In the terms of philosopher Karl Popper, the *Prime Assumption* is therefore an unchecked belief, which might well prove *falsifiable*. As indeed all those pre-call experiments I've reported here now show. Furthermore it's a matter, which others can now always test out for themselves if they so desire.

Confusions clarified

A further advantage of the anti-memory concept, is the large degree to which it can clear up previous confusions in its field. For example, as seen throughout Part Two, it can totally replace - and so make redundant - those former notions of *'clairvoyance', 'telepathy', 'remote-viewing', 'precognition', 'synchronicity', 'ESP'*. Their replacement also affords a large gain in parsimony or mental economy: it empowers us to stop puzzling so much about whatever they might mean.

All these older terms were further derived from faulty or inadequate observation originally – a matter I've also clarified throughout Part Two. Neither did any of them ever succeed in progressing beyond slight initial hints of their reality – which again must suggest that they were all unreal or inadequate to start.

Worse still, these former confusions were always misleading very seriously. For inadequate observation had fatally combined with that common 'space-before-time' habit, to make people see intuition as a space-transgressing faculty – and so one opposed to what the rest of science knows. From which came supposed disharmony between the paranormal and the normal, a spurious conflict that explains the general hostility of so many scientists still.

Such basic conflict was always of course entirely unnecessary, had intuition been interpreted more parsimoniously in terms of just time alone. For the laws of time (assuming of course that there can be any such!) are as yet very much undiscovered and unknown. As indeed the large current uncertainty, about the potential role of space-time in everyday experience, now also shows.

All this supposed conflict, between science and paranormal reality, was therefore entirely unnecessary, and can be traced to elementary errors of observation at the start. As against which, the new anti-memory concept is mostly more harmonious. It seems also capable of near infinite open-ended development - and probably into regions as yet quite unimaginable!

A productive concept

Anti-memory therefore appears to offer much more than a merely correctional or harmonising role. For through direct experiments which were previously unthinkable, it may even enrich conventional science considerably. One obvious example is the current debate over quantum indeterminism, where pragmatic pre-call can now readily falsify that strange and suspect Copenhagen Idea (Ch.16).

Likewise there's the apparent symmetry between past and future, as implied by Relativity Theory (RT). This seems a required conclusion that many physicists have suspected and expressed, but never at all explored pragmatically. But now anti-memory can readily prove its initial everyday reality, though to what extent is still unclear.

Those preliminary findings I've reported here also indicate that both quantum events in micro-time, and everyday events in middle-time, must seem more predetermined than previously realised. As such they conjoin more readily with RT in macro-time, where predetermination is commonly accepted as a large-scale feature reality.

Given such it may follow that the entire spectrum of operational time – from its greatest interval of 15 billion years down to its briefest one of a trillion trillionth of a second (10^{-24} sec.) – might most simply be considered as one great unity overall. If so, this new realisation might suggest a basic unification between RT and quantum theory, an objective greatly desired in physics right now.

Computers can't pre-call!

In similar vein anti-memory can more certainly help to resolve another important issue in current science: is there anything more to Mind or brain than a mere biological machine?

One way to settle this question would be through a decisive Turing Test – as suggested by British pioneer Alan Turing half-a-century ago.[3] He sought some rational criterion through which we might decide, whether computers might ever be said to think as humans do.

To this end we can imagine a human in one room and a computer in another, both linked via terminals to a remote Observer scientist. The issue is then whether this Observer could ever initiate a true Turing Test.

This would be a "conversation", or maybe a series of questions, through which he could identify the human terminal.

So far such a test has proved remarkably difficult to devise. In fact Turing himself suggested that only some manifestation of paranormal function – e.g. 'ESP' - would unerringly reveal the human terminal. His suggestion can now be suitably upgraded through the new understanding which anti-memory can afford.

For example we can now postulate a computer, in competition with a human anti-memory adept at those two remote terminals. Both might be instructed to predict each singular result, in a roulette game controlled by the Observer at all times. In which case, the human should soon pull far ahead in scoring, so readily proving his identity.

For computers could never pre-call by any stretch of the imagination, no matter how programmed. Which therefore highlights a fundamental difference between machine intelligence and our own!

A new kind of consciousness

On a more personal and subjective level, to communicate the nature of anti-memory - or what the actual experience feels like - can indeed be rather difficult. This has to do with what philosophers term *qualia,* our personal experience of reality. For they've long known there's no real method for proving, that the experience of identity is similar in my mind and yours.

For example is your experience of the colour red really similar to mine? We can only agree that it's probably somewhat similar, at least.

To an anti-memory adept, this familiar problem of qualia becomes much magnified. For now to a large extent one is trying to communicate the incommunicable, or something of which the other has little or no conscious experience at all. Nor do we have adequate words to convey what the paranormal experience "really feels like.". The best one can do is to surround the relevant mind-state with partially descriptive adjectives – calm, cool, relaxed, happy, Childlike and so on.

But otherwise to try and communicate the essence of the anti-memory experience is all a bit futile. It's rather like the problem of describing rainbows to the blind. Something essential gets lost with the translation into language during the description process.

On the other hand even a few hours of direct personal experience, can soon give a better impression of what deliberate anti-memory feels like. Such effort also entails deliberate striving towards a new skill of mind, and so a novel state of consciousness. On the possibility of which psychologist William James commented a century ago: [4]

> "Our normal consciousness is but one special type, while all about it, parted by the flimsiest of screens, there lie potential forms of consciousness entirely different"

In more ordinary experience in any case, the deliberate attainment of new states of consciousness, has always been a feature of human development. For example when children first learn to write, count, and so on, their functional consciousness is presumably altered from their earlier state.

Likewise computer programmers can now readily execute mental contortions that would have been quite unimaginable, to their farming grandfathers three generations ago. Presumably therefore their mind-states, or brain contents which empower this new variety of consciousness, must also be to that extent quite different.

But in any case controlled intuition, as attained through deliberate anti-memory, is mostly a pleasant and life-enriching feat of mind. Though at first you may find it to be slightly upsetting or even fearful, that is until you get used to its quirky little ways.

Nor could you ever reasonably aim to pre-call for a full 12 hours daily - no more than you might hope to do advanced mathematics for so long and preserve your sanity. Instead about 1 daily hour is quite sufficient – and preferably focussed on some specific aspect like numbers rather than life in general. Real mediums (and there are a few of them about!), who focus on psi continuously, don't seem to live very happily or normally.

But in any case luckily, controlled anti-memory (in whatever mode) is always a rather labile and delicate mental exercise. So its psychological concomitants are easily suppressed or turned off, and largely forgettable within a week or so. Though then, and again as with other more ordinary skills, they're more readily re-attained after each successive return.

The *plastic* future?

Neither need one let high pre-call competence engender a wider sense of fatalism, through imagined or projected over-extension of its potentialities. Whether the personal future is wholly – as distinct from partially – predetermined is still totally unclear. In fact something like a *plastic* future seems currently more probable. This is a concept explored by English writer J.B.Priestley, who was an avid fan of J.W.Dunne 50 years ago. [5]

It may then be that people all have a *potential* future that would normally transpire into some predetermined version of *actual* reality - providing you don't choose to intervene before it can do so. But through its *plastic* quality, it's merely a potential, or most likely, form of future – and so can always be altered through informed free will. The result may then be that some other potentiality is actualised, and so made *made present or presented,* as you become aware of it.

The fate of some ancient tree in your garden may provide a first example of plastic futures in reality. If left alone, its decaying tissues can predict its *most likely* potential future – i.e. that it will be uprooted by the next good storm. Through free will or choice however, you can always intervene beforehand to produce a *less likely* outcome. For example you might bring in a stout plank to prop it, so ensuring that your ageing tree stays up indefinitely.

Most other everyday futures may likewise be *plastic* - or subject to modification by free will – in similar degree. This might even relate to the problem of psychokinesis (mind over matter) – a paranormal faculty to some extent proven in several computer experiments. With these the data suggests that deliberate application of Mind, may possibly sometimes alter the *most likely* potential future (i.e. computer randomicity) into a *less likely* one (i.e. non-randomicity).

Though here some might argue that you can never alter 'the future', since its definition means merely 'what will be'!

In all such arguments however we very probably need new definitions and terms. For example 'the future' seems a rather nebulous concept overall. Indeed RT suggests that we may all have slightly different futures, dependent of the orientation of our personal world-lines. So that what we call 'the future' may be really an interacting composite of all these personal futures, a reality for which we still have no proper name.

New terms for the temporal would then presumably liberate us, from that old Indo-European straitjacket of inadequate or misleading time words. These are all terms that we've accepted largely unquestioned, from primitive peoples long ago. And through which our notions of 'the future' have rather unthinkingly evolved.

Or as Arthur Koestler expressed this matter, in a slightly different context: [6]

> "We have been ensnared in the logical categories of Greek philosophy, which permeate our vocabulary and concepts, and decide for us what is thinkable and unthinkable"

Finally the historical development of other sciences also affords a valuable lesson here. For, as they developed, they replaced loose common words, with whole glossaries of more specific terms and definitions, which better express their actualities. For example in this way chemistry gained dozens of new terms – element, compound, atom, molecule, bonding, ionic, etc. - as it evolved from the misleading words of alchemy.

In the sphere of time and its properties however, little or no such development has yet occurred. That the language of time is still confused and pre-scientific is generally agreed.[7] This is another way of saying that no true 'Science of Time' has yet evolved.[8]

That too is a very curious omission when you come to think about - an omission that future historians will probably characterise, as one of the greatest defects in our scientific age.

All these are matters one should realise before embarking on any of those various pre-call learning routines.. Still, apart from their scientific emphasis on numeracy, these routines are really little different from those you will find recommended by any number of gurus, or repeated in any number of good self-help intuition books.

In general also numerate routines are preferable to non-numerate ones. For they can provide 'hard' mathematical results. If positive these which will give you more confidence initially, that you are really putting intuition or anti-memory to work.

Technical people should also beware of computerisation at a too early stage. As I've noted in Part Three, there are several large assumptions or unknowns about computer modes. It's also well recognised that pressing keyboard buttons can quickly become addictive. So that it seems best to get in some simple card practice first, before venturing too quickly into less certain electronic fields.

A life-enhancing skill

In any case when approached properly, and with due regard for intellectual balance, the deliberate exercise of anti-memory can soon become a life-enhancing skill. For example it can enrich your social and business life through increased serendipity – the knack of happy accidents.[9] The result can be numerous new contacts and encounters, which might hardly have happened otherwise.

Likewise it can enhance your creativity very obviously, and typically when you're least expecting it.

To develop you latent capability for anti-memory in such ways can also be seen as attainment of new mental power. Einstein is often quoted as claiming that most people only use 10% of their true Mind abilities - though the source and accuracy of this notion is far from clear! It may have had something to do with intuition, towards which he was always very much favourably disposed, as indeed my initial quotation at the start of Part One shows.[10]

Although of course as we've seen already, the great physicist never seems to have investigated much further, into the real nature of what this intuition faculty might be!

But in any case anti-memory development can constitute a new clear step, towards novel mental capabilities. In effect it extends your time awareness, into future realms that seemed previously quite closed. In this way it can be regarded as a new kind of enhanced intelligence – some evidence for which I've sometimes seen.

To anti-memories emerging from the subconscious, may also be ascribed those occasional flashes of sudden illumination, well recognised and prized in the annals of science discovery. Dozens of these 'Eureka Moments' have been recorded; they often result in dramatic leaps of progress, with problems intractable before.

On a humble level one such was the invention of the sewing machine by Elias Howe (1819-67):

> Howe had long been frustrated in his attempts to invent a viable sewing machine. His problem was always how to ensure that the thread would enter into the hole in the cloth.
>
> Then one night he dreamt that he was being assaulted, by fierce savages with spears. The peculiar thing was that these spears had holes drilled through *the pointed end* of them.
>
> Howe rose in the morning and perfected his sewing machine in a matter of days!

On a more elevated level is the well known story of how F.Kekule suddenly dreamt up the ring structure of benzene while dozing by the fire.[11] Another was Irish mathematician William Rowan Hamilton's famed discovery of quaternions while strolling by Dublin's Royal Canal.[12] In my biography of him I've noted how it was said that:

> Often he exhibited an almost uncanny power of intuition, a strange and unusual ability to go straight to the roots of some new problem, though without anything obvious to guide him on the way.

> After which many weeks of the most intense and laboured logic were usually required – to prove that his original hunch was true…

All of which also of course can now make more sense, in terms of anti-memory dimly pre-calling large progress to come. So that any scientist or institute in search of enhanced creativity, might well be advised to acquaint themselves with the subject towards this end!

Inside insight

If strong anti-memory experience then happens to be combined with some science knowledge in a single practitioner, the whole thing becomes much easier to understand. Few mediums or other strong intuitors have ever claimed much acquaintance with science; few scientists have ever claimed much competence with strong intuition skill.

Wherefore when science and intuition are personally combined - as I've continuously reported throughout this book - their productive symbiosis makes it relatively easy to discern what's really going on. This combination I've termed 'Inside Insight'. It resembles having gained entrance through closed borders, into some previously shut mental territory. Whereupon it becomes relatively simple to discern its various realities.

From this viewpoint too the whole field of parapsychology – though not so much the wider field of psychical research - is so full of misunderstandings that it could well merit a full book on its own. For parapsychology remains an enterprise which has made remarkably little progress, despite perhaps 1 million man-hours expended over the past 75 years. From the anti-memory viewpoint, this striking large failure of parapsychology, stems from four clear deficiencies:

First is the failure to check up on those old space transcending notions of intuition, now demonstrably derived from faulty observation of the natural facts initially.

Second is the equally unquestioned adoption of behaviourist routines in the main. These are hardly the most effective way to tackle any problem involving higher mental subtleties – and certainly not the super-subtlety of psi.

Third is a very shallow approach - one that for example equates 100 guesses apiece by 100 students, with 10,000 trials by one motivated individual. This is about as sensible as equating the total of 100 first-day golfers, with the total of some one individual who has been playing for half a year!

Fourth is a total lack of any primary psi experience, among those researchers who aim to study the capability. The whole discipline then rather resembles "blind persons probing rainbows" in this aspect.

All of which further suggests that even a little due personal experience, of intuition in whatever mode, might be a good *sine qua non* for all researchers henceforth. Otherwise it seems hard to see parapsychology ever making much progress, without a reliable source of psi competence always to hand. Just as early electrical research around 1800 required Volta's battery as an 'ever ready' power source, before it could get really under way.

Try it and see!
For those who desire to study intuition optimally, the sensible qualifications therefore now seem quite obvious. At least 100 hours of direct, personal, and successful experience of intuitions – perhaps best in one of those modes I've described here - must seem required.

For that is the very minimum of effort one might expect, to be recognised as half-competent in any other more ordinary intellectual field. Or conversely how could people ever reasonably claim much expertise, in some novel region of which they have no personal experience at all?

It might of course be, that all those experiments and findings I've reported here, are total invention on my part. Happily however that's a presumption inherent with science reports of all kinds, though seldom taken literally. Their usual criterion is therefore what I've previously termed a Galilean one (p. 141). That is to say scientific reports must always embody sufficient instructions, which will enable their audience to "Try it and See!" if so desired.

Will those routines and results I've reported here then prove largely replicable, by others so inclined? *Firstly* I can see no reason why they shouldn't be. For there's nothing strange, magical, miraculous, or otherwise 'psychic', about my own mental makeup as far as can be discerned.

Secondly there are those findings from the remote-viewing program (Ch.9) – by far the most extensive investigation of intuition ever carried out elsewhere. While always mistaken in its focus, the conclusion still was that intuition can always be developed, apparently by almost anyone.

Naturally however some people will always perform better than others - as commonly found with sports, mathematics, music, or indeed any other concentration skill.

Thirdly pre-call learning has often been confirmed by a few others, although admittedly in rather unplanned and haphazard ways. Some of these others were early collaborators on coincidence I've mentioned in Chapter One. More were students who volunteered under a preliminary grant from the New York Parapsychology Foundation in the mid-seventies. Most intensive of all was my colleague Frank Morgan from Atlanta (p. 142) who confirmed much secondary detail.

Wherefore all those who may be interested in anti-memory – be they believers, disbelievers, critics or whatever – are most cordially invited to "Try it and see!" at any time. This they can readily do by following my detailed instructions, in any one of those 5 different pre-call modes I've reported here.

Will my conclusions then further be found reliable and general overall? While I might be wrong in some particulars, I have many good reasons to expect – and few to doubt – that they will be found reliable and replicable in the main. More importantly I believe that those various secondary features – like average scoring 'form' at many points, and temporal inversion with early 3-digit pre-call – will also be found quite general.

Likewise my competence could probably easily be bettered in several regions, for example the non-completed challenge of winning at roulette. For some people are always better than others in any exercise of intellect. And I've never had much reason to imagine, that I should be especially adept in any particular pre-call mode.

Such reasons and experience therefore constitute the confident basis of my invitation to others to "Try it and see!"

A brief summary
The concept of anti-memory fulfils many of the classic functions of good theory – explanation, generality, conformity, productivity, etc. So it can be regarded as more of a theory, instead of a mere hypothesis.

Among its many offerings is a good Turing Test, so indicating that the human Mind/brain must be more than a mere biological machine.

And when due anti-memory experience is combined with some science competence, it becomes fairly easy to grasp its essence, resulting also in heightened creativity.

The entire exercise can also be a fairly demanding one initially. Still there seems to be no good reason, why others should not master this new skill of intellect, and to somewhat higher levels than I've reported here. Its further proof is therefore essentially Galilean, with an open invitation to all comers to "Try it and See!" as so inclined .

21

QUESTIONS ARISING

"There are reasons to believe that the subconscious is ignorant of the temporal aspects of the world: it knows nothing of before and after..."
J. T. Fraser – *The Voices of Time* – 1968 p.588

THAT ANTI-MEMORY PROVIDES a viable new paradigm, for intuitions or psi awareness, seems therefore quite clear. That it requires new understanding, of that personal time which we all experience, seems equally unavoidable. And that new problems and questions are then raised, is an additional consequence.

Still it's in the nature of progress that new theories will raise new problems not previously realised, encountering novel phenomena as they open up fresh territories. For even the most comprehensive new theories seldom solve everything at once.

In this concluding chapter I'll therefore briefly consider a few of these new issues, or questions that arise. These are challenges which must be left to others for further exploration, delineation – and hopefully resolution eventually.

Relativity rules?

Of these basic issues, quite the most fundamental is the "real nature" of time and how we comprehend the same. In this context our common western notion – that time somehow 'flows' as it changes the unformed future into rigid past - was always logically suspect. It's also almost impossible to clarify when you come to think about it all.

But in any case this common view could never accommodate the facts of anti-memory – and to that extent is now falsifiable as well!

Conversely the space-time concept of RT – whereby time and space are intimately intertwined – seems far more hospitable. It offers a more coherent framework within which anti-memory might operate.

On this, as we've seen, many eminent physicists have speculated that our minds are somehow "advancing through laid-out space-time" as it were. And further that we may just grow aware of those future facts which are 'already written' - but were previously hidden from us by the limitations of our consciousness.

Still few or none of these commentators has apparently ever tried to clarify, just what these limitations of consciousness might be. Neither apparently have they ever examined their basic *Prime Assumption* that I've now restated several times – i.e. that Mind or memory can just reflect the past alone. So they never progressed beyond mere speculation, nor proposed any practical experiments to check on that seeming reality implicit in RT.

Einstein's theory however implies that what we term *past* and *future* are largely similar, our purely local labels for these entities being of no real significance. To some truly objective Observer therefore, and especially one with an eye for symmetry, it might almost seem inevitable that anti-memory should occur!

And when such an outlook is further catalysed, by due experience of intuition and understanding of temporal repression, the deliberate development of anti-memory soon turns out to be quite straight-forward and clear. Neither need any of its consequent findings surprise us in the slightest: they're all more or less as RT would apparently suggest!

Past, future, and now

That at least some, of those events that we label *future,* must be more predetermined than previously realised, seems therefore now quite inescapable. For that is the strong implication from RT, and further a consequence that all anti-memory findings now support. What was previously just informed speculation, is thereby now transformed into readily observed fact.

To realise that anti-memory supports past-future similarity, is however far from proving that these two largest time regions must be totally identical. Such would be over-simplistic, and anyway unwarranted at this stage.

Rather the truth may lie somewhere between Einstein and Newton, in some new temporal viewpoint as yet unimaginable. This new viewpoint would have to accommodate both modes of memory, and also the common experience of *now* which is so transient. The core of this problem could be how to reconcile those unattainable (but still knowable) *past* and *future* regions, with *present* experience that is forever being transformed.

So far our best model may be that common illusion of smooth change in the cinema. Another model, which I've suggested, involves a sputtering or stroboscopic fuse of *present* consciousness, one which could change flexible (i.e. *future*) world-lines, into inflexible (i.e. *past*) ones. [1]

New and more adequate word-concepts or definitions, must seem required in any case, here. That the basis of all proper thought is linguistic, is a maxim which applies especially to time. [2]

Rethinking required!

Another common notion, for which anti-memory must demand new understanding or refinement, is the physical doctrine of causality. This requires that there must be a direct link of causation between some event, and another that preceded it. For example the gun only fired *because* the trigger had been pulled.

Likewise we can extend this causal chain into our future experience: the glass will only smash *because* the bullet will hit it just before.

RT however demands that this chain of causation is more properly expressed in a present or tenseless form. So the gun fires within a short space-time interval of the trigger-pulling event; the glass shatters another short interval beyond. The whole causal chain lies timeless in reality; it's only our perception of its past-present-future that can change. All of which of course fits in with the anti-memory experience quite readily.

Critics however may hold that causality is being violated - in a purely mental context – here. They would ague that pre-call implies that some future event has somehow 'reached back' - to cause change in one's brain/Mind and enable the prediction to be made.

Against which three counter-arguments can be advanced:-

1/ To extend purely physical observations, into the largely unknown region of mentalities, must seem unwise or premature at this stage.

2/ If anti-memory were to become firmly established as hard fact, then understanding of causality in this context must inevitably be extended or refined. For reason must always yield to the hard facts of nature, and not the other way around.

3/ The causality argument is in any case dependent on time notions based on *The Prime Assumption*, and as such lacks firm foundation logically....

We can also usefully imagine some far-off planet – perhaps made of anti-matter? – where people enjoy full symmetry of memory. They can consider both past and future equally. Their view of time must then be very different, to the conventional understanding here. Though it would of course be more in keeping with what RT apparently implies.

These time-symmetrical extraterrestrials, would therefore view the entire causal chain of sequence - embodying past, present, future - laid out in one great panoramic unity. And for them the conventional Earth view of causality would describe just one half of reality - with the future half hidden merely because we never thought to investigate the same!

Again too the conventional argument of causality, adopts the common position that something of past experience is (somehow!) stored as memories within the brain. Though here one must note that the actual seat of memory – or just how experience is stored in actuality – has proved remarkably difficult to discern. So that not even the greatest physiologist can yet explain how it works.

It might even be, as Rupert Sheldrake holds, that the brain acts more as a receiver, and not so much a recorder, of our memories. If so, to seek for their storage mechanism might be as futile, as searching for television pictures stored behind the screen! [3]

Nevertheless in the particular case of anti-memory we must clearly distinguish between two possibilities. To investigate the *psychological* similarities – between intuition and past-oriented memory – is clearly a very productive pursuit. But how far this extends into a *physiological* similarity, if indeed at all, must for the moment remain quite unknown.

In any case the anti-memory paradigm doesn't really defy causality as one might initially assume. Rather it just raises the feasibility, of knowing future points as well as past ones, in a causal chain of physical events, that still lie linked and unchanging in space-time. What's really in question here is conventional understanding, of that intimate link between time, memory and consciousness.

And finally in any case the conventional argument, for causality as commonly understood, is now readily falsifiable - by the simple finding that anti-memory works. In which case any arguments against its possibility, are obviously somewhere wrong! As with Galileo's telescope and those who thought that they knew too much to bother looking through it, all arguments can now be subjected, to the hard acid test of direct experiment.

Similar observations also apply to that other common belief in free will, a belief that was in any case never too well defined. Does 'free' mean a choice not determined by anything at all? Or if it is limited by personality factors like habit or disposition; if so, can it really be regarded as 'free'? [4]

QUESTIONS ARISING

Here it seems wisest to adopt the concept of a *plastic* future as a working model meantime. As I've already explained (Ch.20), that would imply a future which would always be predetermined, if we leave things to run their natural course. While still capable of being steered into a different present outcome, through one's freedom of choice or free will deliberately exercised.

In any case anti-memory can open up this entire area to immediate and direct experiment, in a manner unthinkable before. For example one such experiment, which I once conducted briefly and informally, I termed "The Cheating Croupier". This involved a private roulette game where only the croupier could observe the wheel. She was also permitted to 'cheat' – by calling out false results whenever she desired.

The question then was whether she could use her own free will, to negate or neutralise my voiced statements based on anti-memory skill. That she could so with consummate ease hardly needs saying - whereas I could then only judge that she really was 'cheating' overall! Which also suggests that the ever helpful analogy with past-oriented memory, is therefore at best only a partial one.

Neither need anti-memory engender an overly fatalistic outlook, one holding that all our lives are predestined at all future times. On this subject the reader may care to consult philosopher A.J.Flew, on 'Precognition' (sic) in the *Encyclopaedia of Philosophy,* his article being one of the best that I know of on such points. [5]

But here one must also note that all cogitations in this area so far, have always accepted that universal *Prime Assumption* quite uncritically. To that extent they are therefore weakened and suspect. And of course if anti-memory proves generally viable, then the entirety of free will must obviously be rethought anew.

Finally, for what they may be worth, I'll give my own impressions of the anti-memory process as potentially in conflict with free will. Though demonstrably strong in certain well defined modes, (as well as intuitions in general), anti-memory has never seemed to threaten my own impressions of personal free will otherwise. So it has just seemed best to ignore the potential greater implications, proceeding with ordinary life otherwise more or less as usual.

Therefore I still believe, that free will is a valid factor, in the human condition overall. While as yet being unable to fully clarify how this belief can be reconciled, if at all, with the new reality of anti-memory. Perhaps the truth is that free will is somewhat less than people have always imagined, while still not fully absent from our lives and deeds.

Full competence?

A further question, unsettled at the moment, is whether fully reliable anti-memory competence, ever would be possible. To consider this problem, in terms of knowing *everything* that will happen to you tomorrow, must seem far too extreme right now. Rather the question is better reduced, to those 5 modes of numerate pre-call I've described in earlier Chapters. I'll therefore consider just these in more detail,

As I've frequently reported here, any particular act of anti-memory seems always associated with a fairly definite state of mind. In my comments and instructions so far, I've only been able to describe this psi-state through relevant adjectives – clear, cool, happy, etc. But with sufficient perseverance and practice, this special or intuitive mind-state becomes consciously recognisable at most times, though still very problematic to describe.

The question is then whether *most* times could ever be converted into *all* times, with anti-memory competence then always evocable through deliberate full control.

In practice here, for the moment anyway, we can first forget about totally perfect *pre-production* of some very complex future scene. Or in terms more numerate, this might be some number like a Lotto ticket, that could be many digits long.

On the other hand experience shows that just 3 digits should be far more amenable to pre-call skill. Indeed that's the whole substance of my various reports in Chs.13 and 19. The question is therefore whether attempted pre-call, in either 3-digit or the more limited binary mode, could ever be made fully reliable *with every single try?*

On this point my findings are at best inconclusive so far. For example when pre-calling cards in binary modes, the highest scoring level I've ever maintained was around 90%, and then only for a couple of days or three. After that, like some Olympic athlete who can only rise to peak form for a few days, my scoring competence would decline again.

Conversely with car-spotting – which I've never so far pursued to full limits of competence ability - I could blithely go on for hundreds of trials while still getting *at least 1* digit (x-type) correct each time. In this latter activity, as I've previously mentioned, scoring can then be compared to a sort of mental dartboard. On which an expert, even at his most careless, can still always hit somewhere. While not necessarily getting near the centre, or full perfect scoring, every time.

Here too of course those various new marker systems for cerebral physiology – as for example Functional Magnetic Resonance Imaging (fMRI) - might change this situation for the better very readily. Very recent new evidence is now revealing that there is surprisingly little difference in brain activity as people go through two temporally opposed routines – i.e. trying to recall their own past or imagine their future activities.[6]

Which again may be interpreted to support my main contention throughout – that there are strong similarities when memory is operating in either past or future mode.

We may therefore reasonably project that there should again be little difference in fMRI patterns if one were to go the whole hog and extend the imaginative routines fundamentally – with an anti-memory adept changing between re-call and pre-call at will.

If so, this might easily lead to greater precision, by making a pre-call statement only whenever the relevant mental physiology appears strongly on screen. In which case there might be little problem in attaining 100% accuracy at each and every try

Conversely I've sometimes suspected that there might even be some psychological counterpart, to Heisenberg's physical Principle of Uncertainty, in operation here. If so, full certainty at all times might never be attainable. ….

Higher mysteries?

Finally substantiation of anti-memory raises another issue of some concern. So are there really any "higher mysteries" in paranormal awareness, ones that it cannot currently explain?

One such problem sometimes raised is that of *coincident timing* - between some internal intuition and the corresponding external event. For example it's frequently claimed that people may form an intuition about some far-off accident "at the exact moment" it occurred.

But the first fact here is that many of these timings are hardly as 'simultaneous', as people tend to believe. For example those early SPR pioneers soon realised that the coincidence, between a 'crisis moment' and some other person attaining an intuition of it, was seldom exactly synchronised. So they were forced to propose the approximation of *latency* – which meant that timings were not always coincident.

In addition Carl Jung, as we've seen in Ch.8, eventually realised that timing between intuition and reality was more likely to be *synchronic* (around the same time) rather than *synchronous* (at exactly the same time). So that *'Synchronosity'* then became *'Synchronicity'*, in his scheme of things!

All in all therefore the facts suggest, that this common notion of exact timing, is more of a myth than a reality. In popular imagination it's probably catalysed, by that usual habit of thinking in terms of 'space-before-time', and so making easy the acceptance of *telepathy*.

A different problem for anti-memory might be that strange process of D/T 'guessing' initiated by J.B.Rhine. (Ch.7). In this routine the intuitor 'guessed' down through a full pack of at least 25 cards, before he or she got any information at all. Though here, as I've already explained, anti-memory can at least deal with Rhine's finding of *salience*. This was the tendency to score mainly with cards near the start and end of a series.

Otherwise however the requirement must now be for some dedicated archivist to go through all Rhine's guessing records, and separate them out clearly into 3 categories. These would be where the guesser was informed after 25 trials, or after a larger number, or never specifically at all. Anti-memory would then predict a marginal degree of success in the first instance, tailing off to nothing in the third.

. But if there was still significant success throughout, that would entail some higher mystery which anti-memory doesn't explain.

A similar problem for which anti-memory is unfitted would be that of post-death prophecy. Here the example which people usually quote are those 'prophecies' of Nostradamus, supposedly concerned with events long after his death in 1566. But these so-called 'prophecies' are highly vague, ambiguous, unclear. Or if we are to take him seriousl, Nostradamus should have written more clearly throughout.

Otherwise the literature reports very few instances of post-death prophecy at work. Of these the best known is a case I've already reported (Ch.9), wherein remote-viewer Pat Price correctly described a large pressure vessel at Semipalatinsk - a feature not confirmed until several months after he died.

Finally there are those very few most curious cases of coincidence, which anti-memory simply can't explain. Less than 5%, of those incidents that I examined personally (Chs. 1,2), fell into this category. Still some of them seemed most curious, almost awe-inspiring, indeed.

Perhaps some of these formed a hard residue of genuine chance coincidence. But others - like 'The Elusive Butterfly' incident (p.17) - seem almost to have an aura of the supernatural about them still. It must indeed seem rash to imagine that we known all about such matters, or indeed that science can so exclusively forbid them all.

For the present therefore it just seems wisest to just note such potential 'higher mysteries', and then set them aside. They may indeed reflect some higher order of reality which is quite beyond our comprehension at this stage. But meantime more progress is likely to be gained by concentrating on those aspects now obviously amenable.

New theories of course are seldom able to deal with *all* the natural facts initially. And anomalies like these, at first set aside as beyond comprehension or explanation, may eventually start to loom as strong imperfections on the original. Eventually they may even seem so important, as to require some new and better theory to explain. For that is the fate of most paradigms in science, and so possibly too with the whole anti-memory idea.

A final brief summary

I started this book with a personal survey of everyday coincidences, which turned out to be intuitions in the main. They could then be explained in terms of traditional *precognition* – though deeper analysis soon clarified this term as misleading and inaccurate. Instead intuitions are more accurately described as instances of pre-call or anti-memory

This new concept proved further sufficient to deal with all former traditional notions like *remote-viewing, ESP, telepathy,* etc. All of their anecdotes, and at least most experiments, could be explained in terms of anti-memory once again. The new approach also defused that traditional conflict between paranormal experience and normal science. This conflict was ultimately based on unjustified inferences of space violation, arising from faulty observations of intuition originally.

Proof that the anti-memory approach was valid and superior, also emerged with my findings of learned pre-call capability in 5 different modes. These are all cast in a Galilean mode, so enabling all others to "Try it and See!" for themselves.

The three most important pre-call activities are roulette, electrons and car-spotting routines. Overall they suggest that reality at all levels is essentially more predetermined than previously realised.

Substantiation of anti-memory, in so far as I've reported here, seems also seem incompatible with that vague notion that "time passes by!" (ourselves). But this notion is ultimately based on that *Prime Assumption* I've stated in Ch. 4 – i.e. that mind or memory can only contemplate the past alone. An assumption which also lies at the basis of most or all current psychology.

Therein too lies the root of the current great divide between psychology and physics, and their two different views of time. Here the anti-memory viewpoint seems more consonant with RT - and with a surprising degree of concurrence as well.

To know that past and future share a certain similarity from the anti-memory viewpoint, need not however imply that this similarity must be quite total. New words and proper terminology are probably required on such points. Meantime it seems wisest to consider 'the future' as plastic rather than fully predetermined, as indeed my own purely personal experience would further suggest.

Causality likewise is hardly contradicted by such validations of anti-memory. This question reduces to one of our real mental competence and capabilities. And further to what points on the causal chain (which may well be fundamentally timeless) our present consciousness can reach.

Whether full 100% pre-call competence can ever be attained, in even the simplest of situations, is still unclear. So too is the possible reality of 'higher mysteries' – ones which anti-memory can hardly explain. New theories can seldom deal with *all* natural facts initially, and the same is true of the whole anti-memory idea.

APPENDIX ONE

HOW TO GROW MORE INTUITIVE

By now the reader should hopefully realise that intuitions tend to maximise during a rather special state of mind. This special mind-state one can first learn to recognise, then strive to duplicate, and finally reproduce deliberately by controlled exercise of will.

You can further consider this exercise in three ways. Firstly, you can approach it as just a deliberate alteration in your psychological life-style. Secondly you may consider it as cultivation of the Child sector in your personality, the Child being where intuitions are most likely to be found.

Or finally you can think of it as giving more expression to your muted right-brain hemisphere, again the source of intuition and creativity. Then it can more influence your left-brain activities like language, logic, structured thought.

To develop your intuition properly also calls for an in-depth exercise in *phenomenology*. This means deliberate examination, description and replication of your personal mind states.

As yet however there are no words for what you're trying to identify or describe. So you can only try to express, the intuitive mind-state by surrounding it with inadequate adjectives like smooth, calm, relaxed, happy, etc. Likewise it helps to clarify what this mind-state is not. That too can be described by more negative adjectives - like ruffled, excited, harassed, etc..

To keep a special *Intuition Diary* is another good first rule. Preferably you should write it up – even if at first there seems little to record - at some regular hour each day. This will make it a matter of habit and routine, so helping you to develop a growing interest in your project.

Pen-and-paper is also far superior to a computerised word-processor for this task. This is because computers are logical in their input and operation, and so tend to elicit left-brain expression modes. In contrast pen-and-paper are far more personalised, variable and easy to doodle with – and so more generally expressive of right-brain thought.

At first too your early maximisation of the intuitive mind-state - and minimisation of its opposite – can be greatly aided by one of those training routines from Part Three. Since ordinary playing-cards are as good as any, a sensible idea would then be to practice card 'guessing', for about 10 hours over the first 2 weeks. By the end of this period you may still be not very good at pre-calling cards. Nevertheless, because it usually gives correct results, you should have formed a good idea of what the intuitive mind-state (timeless? calm? happy?) means.

Conversely you should be equally familiar with those failure which its opposite (rushed? ruffled? stressed?) entails.

After this set aside one half-hour daily as "Intuition Time" - for deliberate cultivation of the intuitive mind-state. Adopt a restful seating position wherein you can relax readily. Slowly close your eyes and cast out all extraneous 'high-noise' thoughts – irritations, pressures, demands, time-tight schedules.

At the same time keep lazily focused on whatever external issue you may wish to approach intuitively or more creatively. Let your mind drift lazily into a relaxed, carefree, timeless, happy state. Listen mentally for that 'still small voice of calm' which is your Child coming through.

This is all very much as with conventional exercises in meditation or yoga practices. The big difference is that you can now "make more sense" of the whole procedure, through your new understanding of pre-call or pre-memory.

Don't seek out deliberately for new ideas – rather let them come to you. So don't expect novel creativities immediately. At some time over the next week or month is more probable. In any case after your Intuition Time is up take care to revert to your ordinary mental routines. Otherwise it could be tempting - but unproductive! - to continue just drifting mentally all day!

Intuitive insights are also most likely to happen when least expected, often during relaxed periods in your ordinary routines. The main thing is to be sensitive to their faint manifestations, not drowning them out by more strident concerns

Expect also a gradual increase in everyday 'coincidences' (which are mostly real intuitions or manifestations of anti-memory) as you grow better at recreating the intuitive mind-state. Record every such incident, no matter how slight or insignificant it may seem initially, in your Intuition Diary. Note also whatever special factors – e.g. mind-state, time-lag, degree of interest - you may see as associated with the same.

At first you may notice just 1 or less examples of such 'coincidence' each week. But gradually – say over a year or so – this should rise to 1 or 2 daily as your *interest* and intuitive skill grow. By then you can start

APPENDIX

to regard yourself as an experienced intuitor – one with an extra source of intelligence, one which can be turned on and off like a tap at will.

Finally as you grow intuitive at this level, check out how well anti-memory explains your new expertise. Then you will further come to understand how it's always been hidden, by unrealised acceptance of *the Prime Assumption* until now...

APPENDIX TWO:

REPRINT from London ParaScience Conf. – Sept. 1973
A RELATIVISTIC APPPROACH TO SUBJECTIVE TIME
By Sean O'Donnell

It is presently uncertain whether the normal comprehension of subjective time is an irreducibly obvious aspect of the external world (1). Observation however suggests that ordinary temporal notions may owe much to unrealised cultural initiation from infant days. For whereas the newborn seem to arrive with comparatively timeless or even relativistic attitudes, prevalent adult expressions of Newtonian time are then increasingly assimilated over the first 7 years (2).

The derived formal status of time in current psychology is thus seen to assume fundamental principles long found inadequate elsewhere. For whereas a relativistic basis is now unfailingly adopted throughout physical theory, psychological paradigms still implicitly encompass outdated notions of absolute time.

Considering then instead the overall relation between consciousness and the apparent external reality of four-dimensional space-time, one new generalisation is immediately suggested. For it is seen that – with the striking exception of time future – the presently available information content, of increasingly remote space-time regions, decreases by a similar symmetrical process in all four dimensions.

That such is so along any of the six reference directions of the three spatial axes follows as a necessary consequence of the physical inverse square law. That a broadly similar pattern of exponential decrease obtains along the fourth axis of time is an everyday experience quantified in the re-call curves of the Ebbinghaus memory experiments (4).

Diagrammatically the earlier synthesis is then expressed by the general interchangeability, of the purely temporal label **Now-t$'$**, with any of the six spatial labels such as **Here-x$'$**. The relation may equally be summarised in the observation that consciousness exhibits a 7/8 element of qualitative symmetry, in the overall four-fold perspective of external space-time.

DIAG. 1 – similar to Diag. 19.1: this book, p. 196.

When expressed like this however, the missing eighth element of future consciousness is immediately seen to attain outstanding aesthetic and theoretical importance. For were consciousness observed to exhibit some form of symmetrical mirror-memory which could contemplate both past and future with equal facility, a new and currently missing harmony of psycho-physical isomorphism, between present comprehensions of subjective and objective time, would thereby be attained (5).

Such new harmony would result from the general interchangeabilityof the axes depicting subjective spatial and temporal impressions, somewhat as in the objective and non-Euclidean transformations of the Minkowski approach (6). Equally the similar knowability of past and future would indicate their treatment on a less differential basis than before.

Is Memory Necessarily Asymmetric?

Considering then the primal problem of apparently asymmetric memory, careful examination suggests that ordinary notions are not perhaps wholly accurate. This is because occasional time-anomalous experiences - ranging from deja-vu and serendipity to paranormal anecdotes and premonitory dreams – seem to suggest that a sort of future-memory does sometimes occur (7). In all such anomalies the essential common pattern is then seen to be that consciousness seems to contemplate an external event before it is observed, instead of afterwards as in the more normal experience.

It is similarly clear that common parapsychological notions of transspatial processes, such as telepathy or ESP, are scientifically superfluous to first-order explanation. Economy of hypothesis demands that the real-life anecdotes, from which such notions were originally derived, be more simply construed as anomalies of subjective time alone. For it is a necessary common condition in nearly all such tales that the narrator observe the later external complement to his earlier internal thought. And were this not so, he could hardly have any paranormal tale to tell.

The single essential anomaly of time inversion alone then suggests that most such experiences are most simply described by the new and self-explanatory term 'pre-call'. In contrast to a typical three acts of recall for every lived second, the average individual may be allowed at least one such experience at some point in life. The pre-call/re-call abun-

APPENDIX

dance ratio is thus seem to assume a maximum lower limit around $<10^{-10}$.

Given then that the past-asymmetry of memory might not necessarily be total, examination of the comparatively great imbalance between its two apparent modes is next required. For – common and ill-informed temporal notions apart – no such imbalance seems inherent in the relativistic equivalences between space and time, or past and future, in the Minkowski approach.

With attention thus directed towards the innate capacities of consciousness and the normal re-call process, it is immediately seen that the everyday notion of memory, as necessarily past-asymmetric, is no more than an unchecked assumption at best. For if the growing comprehension of subjective time by the learning infant can be correlated with increasing facility in normal re-call, the generally suspect comprehension of the former is most simply traced to accidental bias in the latter.

DIAG. 2 – Similar to Diag. 4.4: this book, p. 35.

It may thus be deduced that consciousness might possess an inherent but latent capacity for direct non-inferential future awareness. If so the sporadic and infrequent modes of its apparent manifestations suggest strong subconscious censorship most simply ascribed to basic disharmony with the rest of the learned world-picture. Man's apparent capacity for occasional future awareness thus seems most closely described in terms of an undeveloped faculty of repressed pre-call.

It may equally be seen that current psychology is unable to accommodate many parapsychological findings only because both approaches are ultimately based on the outdated paradigm of Newtonian absolute time. The new relativistic viewpoint is however capable of containing both disciplines, while simultaneously harmonising currently disjunctive interpretations of subjective and objective time.

It is similarly relevant that no assumption is scientifically acceptable without some measure of investigative validity. Nevertheless nobody ever seems to have checked on the universal and wholly primal assumption that memory is necessarily past-asymmetric. In common with everyday notions of subjective time, all derived psychological, philosophical and logical systems are then seen to be based on a notably unchecked and possibly invalid premise.

(There followed 12 numerate reports on experimental learned pre-call, much as this book now recounts in Parts 3 and 4....)

1/ G.J. Whitrow – *Nat. Phil. Of Time* – London 1961 – p. 51
2/ J.E. Orme – *Time, Experience, Behaviour* – 1969 – p.44
3/ *ibid* – p.181
4/ H. Ebbinghaus - *Uber das Gedachtnis* – Berlin 1885
5/ J.T. Fraser – *The Voices of Time* – London 1968 - p.217
6/ H.Minkowski et al. – *The Principle of Relativity* – 1923 - p.76
7/ W.Goody – *Individual Psychology* – p.83, May 1959

NOTES

Ch.1
1 – Dunne's work is described more fully in Ch.10
2 – These collections of anecdotes are detailed in PART TWO
3 – Plimmer, M, King, B *Beyond Coincidence* - 2006
4 – This notion is considered more fully in Ch.7
5 – The *OED* likewise defines intuition in two ways:
A/ immediate apprehension by mind without the intervention of any reasoning process.
B/ any particular act of such apprehension
6 – Thouless, R – *Experimental Psychical Research,* 1963 p.142
7 – Vygotsky, L – *Thought and Language* – MIT – 1962

Ch.2
1 – Dunne, J – *An Experiment with Time* – 1927 – Ch.7
2 – Husserl, E – *Logical Investigations* - 1900
3 – Sinclair, U – *Mental Radio* – 1934
4 – Gallwey, T – *The Inner Game of Golf* – Cape - 1981
5 – Freud, S – *Collected Papers* – 1955 – v.18 – p.217
6 - He had a habit of washing his hands compulsively or symbolically, and a local nickname which described the same.

Ch.4
1 – de Pablos, F – *J.Soc.Psychical Res.* – Oct. 2004 – p. 226
2 – Augustine, St – *Confessions* – Dorset - 1961 – Ch.10
3 – Russell, B – *Mysticism and Logic* – 1917 – p.202
4 – Ayer, A J – *The Problem of Knowledge* – 1956 - p. 166
5 – Broad, C – *Forekenowledge* – *Encyc. Phil.* – v.6 - p.438
6 – O'Donnell, S - *New Scientist* - Feb 8, 1973 - p. 329
7 – Since the nervous transmission rate is ca. 100 ms⁻, a stimulus may take about .02 seconds to travel from toe to brain.
8 – Vygotsky, L. – *Thought and Language* – MIT - 1962
9 – Orwell, G – *Collection of Essays* –1954 – p.162
10 – O'Donnell, S – *I. Conf. Parapsychology* - poster– 1985
11 – The term 'anti' came into science with P. Dirac's description of the 'anti-electron' (i.e. positron) in 1931.
12 – Myers, F – *Proc.S.P.R.* – 1894 - p.404

13 – O'Donnell, S. – *Parascience Conf.* – London – Sept. 1973
14 – Ebbinghaus, H – *About Memory* – 1882
15 – Janus was also the deity of *janitors* – people who look in and out from doorways!
16 – Stevenson, I - *J. American Soc Psy. Res.* – 1960 - v.54
17 – O'Donnell, S – *William Rowan Hamilton* - 1983 - p.180

Ch.5

1 – Synge, J – *New Scientist* - 1959 - p.410
He was nephew of famed Irish Playwright J.M Synge who wrote *The Playboy of the Western World* in 1907.
2 – Whitrow, G J - *Time in History* – OUP – 1985 - p.5
3 – *ibid - The Natural Philosophy of Time* – 1981 – Ch.1
4 – Bishop, P - *Intellectual Intuition in Kant, Swedenborg* 2,000
5 – Forrest, D - *Hypnotism* - 1999 – p.72
6 - de Puysegur, C – *Animal Magnetism* – Paris - 1784
7 – See *Oxford English Dictionary*
8 – Petetin, Dr – *Animal Electricity* – Paris - 1808

Ch.6

1 – Crookes, W - *Quarterly J. Science*- July 1871. The French astronomer Camille Flammarion had earlier proposed this term.
2 – Richet,C - *Thirty Years of Psychical Research* - 1923
3 – Myers, F W – *Proc. S.P.R.* – 1 – 11 – p.147
4 – Myers, F W - *Human Personality* – 1903 – p.288
5 – Myers, F W, et al. - *Phantasms of the Living* – 1886
6 – *Proc. S.P.R.* – Vol 39 – p.212
7 – West, D J – *Psychical Research Today* - 1962 – p. 162
8 – Sheldrake, R - *J.Soc. Psy. Res.* – 67 - p.184
9 – A letter from me on these lines, in Feb. 2004 to the Editor *J.S.P.R.*, was not acknowledged initially, nor further discussed, nor ever published.
10 – Sinclair, U – *Mental Radio* - 1934
11 – Sherman, H – *Thoughts Across Space* – N.Y. - 1970

Ch.7

1 – Sudre, R – Treatise of Parapsychology – 1961
2 – Rhine, J B & Pratt, J G - *Parapsychology* – Blackwell – p.57
3 – Rhine, J B – *Extra-Sensory Perception* – Faber - 1935 - p.83
4 – Heywood, R – *The Sixth Sense* – Pan - 1966 – p.156
5 – Rhine, J, B, – *The reach of the Mind* – Pelican 1954 – p.127
6 – Rhine, J B & Pratt J G – *ibid* – p. 48
7 – Rhine, L – *Hidden Channels of the Mind* – 1962

NOTES

  8 – e.g. – *J. S.P.R.* – Jan. 2005 - p. 18
  9 – Thouless, R – *Experimental Psychical Research,* 1963, p.142
  10 – Radin, D – *The Conscious Universe* – Harper – 1997
  11 – Robert, R – *Parapsychology* – 2001 – Ch.7

Ch.8
  1 – Bishop, P – *Intellectual Intuition in Kant, Swedenborg,* 2000
  2 – Koestler, A - *The Roots of Coincidence* - 1972
  3 – Fraser, J T – *The Voices of Time* – 1968 - p.222
  4 – I have been unable to trace the source of this anecdote.
  5 - *ibid* – p. 227

Ch.9
  1 - ww.biominds@superpowers.com
  2 – Targ, R & Harari, K – *The Mind Race* – 1986 – p. 13
  3 – McMoneagle, J - *Mind Trek* – 2000 - p.46
  4 – Graff, D – *Tracks in the Wilderness* - 2003 - p.117
  5 – Gruber, E – *Psychic Wars* - 1999 - p. 16
  6 – Morehouse, D - *Psychic Warrior* – 1996
  7 – Abundant material about AIR is now on the Internet.
  8 – McMoneagle, J – *Remote Viewing Secrets* – 2,000 - p153

Ch.10
  1 – Whitrow, G J – *Natural Philosophy of Time* – 1981 – p. 365
  2 - de Pablos, F – *J. Soc. Psy. Res.* – v.68.4 – Oct. 2004
  3 – Barker, J C – *London Med. News Tribune* – Jan. 20, 1967

Ch.11
  1 – Piaget, J – *The Construction of Reality in the Child* - 1954
  2 – Synge, J – *New Scientist* – 1959 - p.410
  4 - Freud, S – *Collected Papers* - v.4 – 1959 – p.368
  4 – Jung C G – *Synchronicity* – 1954 – p. 43
  5 – Berne, E – *Games People Play* – 1966
  6 – Ornstein, R – *Psychology of Consciousness* – 1972 - p.55
  7 – This section was refused a hearing, with no reasons given, when proposed for the Annual SPR Conference - 2006

Ch.12
  1 – Attempts at learned guessing by others began this year, with about 20 such attempts over the next 10 years. Cf. 7/ below.
  2 - Bacon, F – *Novum Organum* - 1620
  3 – Husserl, E – *Logical Investigations* - 1900
  4 – B.F. Skinner was the doyen of behaviourists around the middle of the 20th century.
  5 – Koestler A - *The Sleepwalkers* - 1959 – p.368

6 – O'Donnell, S - *Parapsychology Review* – May 1974
7 – Tart, C – *Learning to Use ESP* – 1975 – p.49
8 – The main exception to this generalisation was an extensive experiment by H. Schmidt, considered further in 16-8 below
9 – Hardy, A – *The Challenge of Chance* – 1973

Ch.13
1 – *Rand Corporation* - A Million Random Digits – *1963*
2 – In 1975 Frank Morgan of Atlanta published the first booklet, which assembled my pre-call publications to that point.
3 - Tart, C – *ibid* – p. 76
4 – Dunne, J W – *An Experiment with Time* – 2001 – p.25

Ch.14
1 – Bass, T A – *The Newtonian Casino* - 1985 – p.119
2 – O'Donnell, S – *William Rowan Hamiltom* - 1983
3 – Tyrrell, G N M – *The Personality of Man* – 1948 – p. 21
4 – Berne, E - *Games People Play* - 1966

Ch.16
1 - Whitaker, A – *Einstein, Bohr and the Quantum* - 1996
2 - Gribbin, J *In Search of Schrodinger's Cat* – Bantam - 1984
3 - Hoyle, F – *New Scientist* – Sept. 10, 1994
4 - Albert, D – *Scientific American* – May 1994
5 - Theocratis, T & Psimopoulos, M – *Nature* – June 2, 1988
6 - Casti, J *Paradigms Lost* – 1992 – Sec. 7
7 - Barry, P – *New Scientist* - 2006 – Sept. 30
8 - Bass, T – *The Newtonian Casino* – Penguin - 1991
9 - Schmidt, H – *J. Parapsychology* – Vol.33, 2, 99-108
10 - Charlie Wells was a Glasgow swindler, who had an anomalous, and likely paranormal, run of luck at Monte Carlo in 1891.
11 - *Science* - July 30 – 1982

Ch.17
1 – See articles on *Being* and *Change* in *Encyc. Of Philosophy*
2 – A country-and-western song of this title was very popular in the early 1960s, featuring also in a Marilyn Monroe film.
3 – Fraser, J - *The Voices of Time* – 1968 - p. 18
4 – Ayer, A J – *The Problem of Knowledge* – 1956, p. 166

Ch.18
1 – Whitrow, G J – *Natural Philosophy of Time* – 1981 - p.274
2 – Einstein's two theories of relativity – the *Special* and the *General* – dealt with velocity and acceleration respectively.

3 – Cf. Ch.11 –1
4 –Eddington, A - *Nature of the Physical World* – 1929 – Ch.4
5 – Whitrow, G J – *ibid* – Ch.5
6 – Davies, P - *About Time* - 1988 – p.283
7 – Weyl, H – *Phil. Of Maths. And Nat. Sciences* – 1949 – p.116
8 – de Beauregaard, C – *The Voices of Time* – 1968 – p. 417
9– Feinberg, G – *Quantum Physics and Parapsychology* – 1974
10– Williams, D – *J. Philosophy* – 1951 – p. 457
11 – Davies, P - *God and the New Physics* – 1990
12 - Penrose, R. - *The Emperor's New Mind* - 1990 – p.574
13 – *ibid* - p.394
14 - Hawking, S – *A Brief History of Time* – 1988 – p.144

Ch.19
1 – *Dictionary of Psychology* – 1999 – p.1066
2 – Ebbinghaus, H – *About Memory* - 1882
3 – O'Donnell, S - *Parascience Conference* – London – 1973
4 – Dirac, P - *Proc. Royal Society* - 1931
5 – Fraser, J T – *The Voices of Time* – 1968 – p.217
6 – Penrose, R – *ibid*. – Ref 18.14
7 – Synge, J L – *New Scientist* – 1959 – p.410
8 – I adapted this term from 'train-spotting' – a hobby where British enthusiasts attend at railway stations to note the numbers of engines and trains.
9 – In classical lore the Fates were 3 sisters who spun out the web of human lives, decided their length, and cut the cloth accordingly!

Ch.20
1 – Gell-Mann, M – *The Quark and the Jaguar* – 1994 –p.75
2 – Kuhn, T – *Structure of Scientific Revolutions* – 1962 – p. 52
3 – Turing, A - *Mind,* 59, 1950, No. 236
4 – Tyrrell, G N M – *The Personality of Man* – 1948 – p.43
5 – Priestley, J.B. – *Man and Time* – Aldus - 1964
6 – Koestler, A. - *The Roots of Coincidence* – 1972 - p.100
7 – Benjamin, C - *The Voices of Time* – 1968 - p. 6
8 – O'Donnell, S – *The Science of Time* – Technology Ireland – Feb. 1975
9 – 'Serendipity' derives from a fabled king of Serendipe (Ceylon) who had the happy knack of doing good by coincidence.
10 – Cf: Introductory quote, Part One
11 – Crump, T – *A Brief History of Science* – 2002 – p. 187
12 - O'Donnell, S – *William Rowan Hamilton* – 1983, p.125

Ch. 21
1 - O'Donnell, S - *Future Memory and Time* - 1996 – p. 184
2 – Vygotsky, L – *Thought and Language* – M.I.T. - 1962
3 – Sheldrake, R – *The Presence of the Past* – 1986 – p.168
4 – Tyrrell, G.N.M. - *The Personality of Man* – 1948 – p.95
5 - Flew, A J - *Encylopaedia of Philosophy*
6 – Marshall, J. *Memories of the Future* – New Scientist 2007 – March 24

INDEX OF NAMES

Aberfan disaster	92	de Puysegur C	37-44
AIR report	80	Dostoevsky F	143-7
Alcock JE	109,120	Dunne JW	13, 85-94, 107
Anaximander	166	Ebbinghaus H	32, 125, 132
Aquinas T	172		186
Augustine St	25	Einstein A	1, 103, 133,
Ayer AJ	26		173-84
		Emerson RW	43
Bacon F	39, 110	Euclid	38
Barker J	93		
Barna Woods	8	Feinberg G	180
Barrett W	46	Feynman R	11, 99, 157. 180
Berne E	102, 140	Flew AJ	216
Bergson H	68	Fraser JT	163, 187, 213
Bohr N	153-63	Freud S	15, 101
Bohm D	156		
Broad, CD	26, 88	Galileo G	116
Bronowski J	125	Gallwey T	14
		Galway	4, 195
Carlyle L	43	Gell-Mann M	95, 199
Carroll L	23	Gladstone W	45
Casti J	153-7	Godley J	90-93
Chomski N	101	Graff D	75-8
Copenhagen Idea	154-64	Gurney E	47
Craig M	14, 52, 63, 99		
Cramer J	156	Hamilton WR	136, 207
Crookes W	45	Hammid H	76-80
Curie M	34	Harari K	73-5
		Hardy A	118
Davies P	165, 179-82, 185	Hawking S	38, 182
de Beauregaard C	171, 180	Heisenberg W	153, 219
de Broglie L	155, 197	Heraclitus	168, 174
de Fermat P	117	Honorton C	63
de Pablos, F.	24, 89,98	Howe E	207

Hoyle F	155	Orwell, G	27
Husserl E	13		
Huxley T	101	Parapsychol. Foundation	210
Hyman R	80	Parascience Conf.	186, 227
		Pascal B	117, 134
James W	172, 204	Pauli W	67-72, 9, 187
Janus	32	Pearce H	55
Johnson M	51	Penrose R	181
Jung, C.G.	67-72, 97, 101, 187	Petetin D	42
		Phantasms	47-9, 61
		Piaget J	100, 187
Kant I	3, 39	Podmore F	47
Kekule F	207	Popper K	158, 201
Kelvin W	137	Price P	76, 220
Kepler J	68	Priestley JB	88, 205
Koestler A	67-72, 205	Puthoff H	74-9
Kuhn T	200		
		Race V	41-4
Linzmayer A	58-60	Radin D	64
Lodge O	45-7	Richet C	45, 55
		Rhine, JB	55-66, 97, 118, 133
Maugham S	146	Rhine L	62
Maxwell JC	157	Russell B	26
May E	80		
McDougall W	55	Schmidt H	159
McMoneagle J	75-83	Schrodinger E	153-6
Medawar P	199	Shakespeare W	113
Merlin Park	8	Sheldrake R	51, 212
Mesmer F	40-4	Sherman H	53
Michelson/Morley	34	Sinclair U	14, 75
Minkowski H	30, 176-84	Skinner BF	113
Morgan F	130, 196, 210	Smart P	51
Murray G	13, 52, 67, 99	Sperry R	102
Myers, F W	28, 47-9	Spinoza B	35
		S.P.R.	28, 46-54, 96
NASA	38	Stuart C.E.	55, 591
Newton I	41, 168-73	Sudre R	56
Nostradamus	216	Swann I	73-85
		Swedenborg E	39, 73, 87
Ockham's Rule	19-24, 57, 62	Synge JL	38, 96, 187

Targ R	73-8	Walpole H	43
Tart C	116-9, 130	Warcollier R	75
Thompson E	78	Watts C	64
Thouless R.	10, 63	Weber-Fechner Law	186
Titanic	33	Weyl H	180
Turing A	202	Wheeler J	157
Twain M	49	Whitrow GJ	38, 88, 96
Tyrrell GN	50, 137	Williams D	181
		Zener D	57
Utts J	80	Zeno	171-4

"There is nothing more difficult to take in hand, more perilous to conduct, or more uncertain in its success, than to take the lead in the introduction of a new order of things.

"Because the innovator has for enemies all who have done well under the old conditions, and but lukewarm defenders in those who may do well under the new....".

N.Machiavelli, - *The Prince* – 1513